Integrated Navigation and Guidance Systems

Integrated Navigation and Guidance Systems

Daniel J. Biezad
QED Educational Services

EDUCATION SERIES
J. S. Przemieniecki
Series Editor-in-Chief
Air Force Institute of Technology
Wright–Patterson Air Force Base, Ohio

Published by
American Institute of Aeronautics and Astronautics, Inc.
1801 Alexander Bell Drive, Reston, VA 20191

MATLAB™ is a registered trademark of The MathWorks, Inc.

American Institute of Aeronautics and Astronautics, Inc., Reston, Virginia

1 2 3 4 5

Library of Congress Cataloging-in-Publication Data

Biezad, Daniel J., 1944–
    Integrated navigation and guidance systems / Daniel J. Biezad
       p. cm.—(AIAA education series)
    Includes bibliographical references and index.
    1. Aids to air navigation. 2. Guidance systems (Flight)
I. Title   II. Series.
TL695.B54   1999
629.132′51–dc21   98-54413
ISBN 1-56347-291-0 (alk. paper)

Copyright © 1999 by the American Institute of Aeronautics and Astronautics, Inc. All rights reserved. Printed in the United States of America. No part of this publication may be reproduced, distributed, or transmitted, in any form or by any means, or stored in a data base or retrieval system, without the prior written permission of the publisher.

Data and information appearing in this book are for informational purposes only. AIAA is not responsible for any injury or damage resulting from use or reliance, nor does AIAA warrant that use or reliance will be free from privately owned rights.

AIAA Education Series

Editor-in-Chief
John S. Przemieniecki
Air Force Institute of Technology (retired)

Editorial Board

Earl H. Dowell
Duke University

Michael L. Smith
U.S. Air Force Academy

Eric J. Jumper
University of Notre Dame

Peter J. Turchi
Ohio State University

Robert G. Loewy
Georgia Institute of Technology

David M. Van Wie
Johns Hopkins University

Michael N. Mohaghegh
The Boeing Company

Anthony J. Vizzini
University of Maryland

Conrad F. Newberry
Naval Postgraduate School

Jerry Wallick
Institute for Defense Analysis

Terrence A. Weisshaar
Purdue University

Foreword

Integrated Navigation and Guidance Systems textbook by Daniel J. Biezad is a comprehensive exposition of the modern navigation and guidance sysems for flight vehicles. It represents a novel approach to teaching an aeronautical engineering course as a comprehensive and integrated exposure to several interrelated disciplines combined with laboratory computer exercises for the students. The text reflects years of teaching experience of the author, while with the Air Force Institute of Technology and California Polytechnic State University, and his pioneering work on the Global Positioning System while on active duty with the U.S. Air Force.

Chapter 1 presents the basic principles of navigation and the concept of an integrated navigation system. Chapter 2 discusses Newton's laws applied to navigation, including the concept of geodesics, basic reference frames, simplified aerospace vehicle equations, and the two types of navigation: Inertial Navigation System (INS) and Global Positioning System (GPS). Chapter 4 introduces the concept of uncertainty in navigation, INS error propagation, probabilities, autocorrelation, and the method of least squares. Chapter 5 discusses practical Kalman filters and their key role in the integration of aircraft avionics systems. Chapters 6 and 7 provide a complete description of the underlying theory for the global positioning system and its application to navigation, including a discussion of the system accuracy. Chapter 8 deals with GPS applications to precision approach and landing, attitude control, and air traffic control. Chapter 9 describes flight testing of navigation systems. Finally, Chapter 10 provides computer exercises for error analysis and Kalman filter simulation using the software provided with this text. The software is provided as single, self-extracting file executable from either the Windows or DOS modes.

The Education Series of textbooks and monographs published by the American Institute of Aeronautics and Astronautics embraces a broad spectrum of theory and application of different disciplines in aeronautics and astronautics, including aerospace design practice. The series also includes texts on defense science, engineering, and management. The complete list of textbooks published in the series (over sixty titles) can be found on the end pages of this volume. The series serves as teaching texts as well as reference materials for practicing engineers, scientists, and managers.

J. S. Przemieniecki
Editor-in-Chief
AIAA Education Series

to Karen

Table of Contents

Preface xi
Acknowledgments xiii
Nomenclature xv

Chapter 1. Navigation over Earth's Surface 1
  1.1 Basic Principles of Navigation and Position Fixing 1
  1.2 Air Navigation Radio Aids 5
  1.3 Flight Management for Situational Awareness 10
  1.4 Closure 13

Chapter 2. Newton's Laws Applied to Navigation 15
  2.1 Math Tools 16
  2.2 Geodetics and Basic Reference Frames 20
  2.3 Simplified Aerospace Vehicle Equations 25
  2.4 Fundamental Navigation Equation 29
  2.5 Inertial Navigation System Solution—A Preview 32
  2.6 Global Positioning System Solution—A Preview 35
  2.7 Closure 38

Chapter 3. Inertial Navigation Sensors and Systems 39
  3.1 Aircraft Gyroscopic Flight Instruments 40
  3.2 Inertial Instrumentation 45
  3.3 Inertial Platform Mechanizations 53
  3.4 Attitude-Heading Reference Systems 70
  3.5 Closure 71

Chapter 4. Concept of Uncertainty in Navigation 73
  4.1 Inertial Navigation System Error Propagation Equations 74
  4.2 Probability Concepts in Navigation 82
  4.3 Least Squares 91
  4.4 Closure 93

Chapter 5. Kalman Filter Inertial Navigation System Flight Applications 95
  5.1 Preliminaries 96
  5.2 Kalman Filter: Discrete Case Derivation 97
  5.3 Kalman Filter Divergence 101

5.4	Aided Inertial Navigation System and the Inertial Navigation System Error Budget	105
5.5	Closure	107

Chapter 6. Global Positioning System **109**
  6.1 Principles of Operation 110
  6.2 Global Positioning System Segments 122
  6.3 Selective Availability 132
  6.4 Global Positioning System Error Sources and Error Modeling.... 134
  6.5 International Satellite Navigation Systems 135
  6.6 Closure 137

Chapter 7. High Accuracy Navigation Using Global Positioning System **139**
  7.1 Global Positioning System Aiding Inertial Navigation Systems ... 140
  7.2 Differential Global Positioning System 143
  7.3 Wide Area Augmentation 146
  7.4 Closure 148

Chapter 8. Differential and Carrier Tracking Global Positioning System Applications **151**
  8.1 Review of the Kinematic Carrier-Phase Methodology 151
  8.2 Aircraft Precision Approach and Landing Using Differential Global Positioning Systems 156
  8.3 Aircraft and Spacecraft Attitude Control 160
  8.4 Air Traffic Control 161
  8.5 Test Ranges 162
  8.6 Closure 164

Chapter 9. Flight Testing Navigation Systems **165**
  9.1 Testing Autonomous Inertial Systems 165
  9.2 Inertial Navigation System Flight Tests 165
  9.3 Testing Integrated Navigation Systems 176
  9.4 Closure 178

Chapter 10. Navigation Computer Simulations **181**
  10.1 Installation of AINSBOOK on Your Personal Computer 181
  10.2 AINSBOOK Inertial Navigation System Error Analysis Simulation 182
  10.3 AINSBOOK Kalman Filter Simulation 193
  10.4 Closure 206

Appendix A Abbreviation and Acronyms **209**
Appendix B Discussion Questions for Integrated Navigation and Guidance Systems **213**
Appendix C Web Sites by Chapter **219**
References **221**
Index **227**

Preface

"The great ocean of truth lay all undiscovered before me."
                                                                    Isaac Newton

Navigation is an ancient art. The earliest mariners followed coastlines, discovered they could use the stars, and took the ultimate risk as explorers in braving the open sea. Ancient Phoenicians navigated with the North Star as a compass. Today we navigate in space using the same heavens that Odysseus did when he kept the Great Bear on his left during his travels.

With the advent of the magnetic compass, sextant, and accurate chronometers for time, it was possible to determine accurately both latitude and longitude at sea. The story of the development of accurate timepieces is especially fascinating. A British cabinetmaker named John Harrison is recognized today as the genius behind accurate time keeping, an advantage that England used to establish its hegemony at sea. It is in recognition of this accomplishment that the relatively arbitrary point of zero longitude is today internationally established at Greenwich, England, and the international time standard is Greenwich mean time (GMT). Accurate clocks caused a revolution in maritime history that extends to this day.

Similar revolutions occurred in air navigation. In the early twentieth century, several radio-based navigation systems were developed. RADAR caused a revolution in navigation technology and has been widely used since World War II. Inertial guidance, thought to be impossible by many scientists because of its unstable basis in mathematical theory, was made practical by Charles Draper and is today the backbone of military navigation and weapons delivery systems.

Today we are in the midst of a new technological revolution—the Global Positioning System (GPS). Starting as Air Force System 621B (as a follow up to the existing Navy system called TRANSIT), GPS has evolved into a system that solves the problem of obtaining accurate position, velocity, and time (PVT) regardless of location, weather, or time of day. GPS takes advantage of using digital pseudorandom noise (PRN) codes to establish both position and time. Since GPS signals are digital and are transmitted using spread-spectrum technology, the system is relatively reliable and jam-resistant. The civilian market interest in GPS has been explosive and shows no sign of slowing down.

This manuscript is written from an operational as well as an analytical point of view. It describes both the underlying principles and operational implications of avionics systems and displays as they apply to practical navigation in the air. The goal is to provide a "jump start" for those involved in developing, testing, or operating advanced integrated navigation systems and to provide an appreciation of the fascinating technology underlying their use. It is divided into 10 chapters: the first five covering traditional ground-based radio aids to navigation, inertial guidance, and Kalman filtering; and the second five primarily devoted to GPS, its integration with other systems, and its applications within the aviation community. The fundamental mathematical basis of navigation is presented, but applications are emphasized over abstract derivations.

The material as textbook is suitable as a tutorial and interactive problem set for senior-level and above engineering students in an aerospace curriculum. A comprehensive set of exercises in the final chapter of the text will run on any PC system with a DOS or Microsoft Windows operating system without additional software. The detailed exercises allow interaction with the complex equations and filtering processes to provide both insight and a rewarding first experience with navigation systems simulation and analysis.

The notation evolved from my initial exposure to the excellent inertial navigation texts in the 1960s and 1970s from the Massachusetts Institute of Technology (MIT). Of course other schemes for coordinate transformations and reference frame identification are quite ingenious, as can be seen by the superb text by Stan Schmidt that accompanies the software. However, after a decade of teaching the material, I have settled on a system that seems to provide the least confusion for those learning the material the first time. As an example, the relative rotation of reference frame C_b with respect to frame C_i is defined by $\omega_{b/i}$ with a cross-product rotation matrix $\Omega_{b/i}$. If expressed in C_b coordinates, the vector and matrix are $\omega_{b/i,b}$ and $\Omega_{b/i,b}$, respectively. The coordinate transformation matrix from C_i into C_b is C_i^b. Thus, a comma in any subscript indicates that a coordinate frame designation will follow. The symbols in this fashion remain relatively clear even in long, complex equations or when used with nested exponents.

Daniel J. Biezad
January 1999

Acknowledgments

I am especially grateful to Stan Schmidt for his practical insights into inertial navigation principles and Kalman filtering techniques that are covered in this book and in the laboratory exercises that accompany it. I also would like to acknowledge the positive encouragement I received from John Pzermeniecki, a talented author in his own right, and from Dan Gleason, who commissioned the initial idea for this work at the United States Air Force Test Pilot School, Edwards Air Force Base, California. Finally, I thank Gayle Downs for her creative efforts on the illustrations and Candice Chan for her meticulous editing.

Nomenclature

A	= system dynamics matrix
A_k	= discretized system dynamics matrix at t_k
a_{sf}	= applied specific acceleration or applied specific force
B	= relative bearing (rad)
B	= system control matrix
B_k	= discretized control matrix assuming control constant over interval Δt
C_g	= gyro damping coefficient
C_v^b	= direction cosine matrix from C_v coordinates to C_b coordinates $[C_3^b(\phi)][C_2^3(\theta)][C_v^2(\psi)]$
C_b	= coordinate system for vehicle body axes with origin at vehicle mass center
C_c	= inertial navigation system (INS) computer-calculated reference axes (in the presence of position error)
C_{cv}	= computed local-level frame
C_d	= principle data axes
C_e	= Earth surface-fixed axes (North-East-Down relative to ellipsoid, commonly called the *tangent* frame)
C_{ECEF}	= coordinate system for Earth-centered, Earth-fixed axes.
C_I	= inertial axes of a Galilean nonrotating reference frame
C_i	= inertial frame nonrotating with respect to the stars at Earth's center "io"
C_n	= navigation axes
C_p	= INS platform reference axes
C_t	= true, error-free idealized reference frame
C_v	= vehicle-carried-vertical axes (North-East-Down relative to ellipsoid)
d	= distance, nm or degrees of great circle arc
f	= flattening of the earth
f_{AERO}	= aeronautical atmospheric forces acting on an aircraft (lift-drag-side force)
$f_{G.v}$	= force of gravitational mass attraction at origin of C_v frame in C_v coordinates
$f_{g.v}$	= "local" gravity force (plumb-bob) at origin of C_v frame in C_v coordinates
f_{local}	= local force [$f_{local} = f_g + f_{AERO} + f_{NGA}$ where $f_g = f_G - \Omega_{(e/i)}\Omega_{(e/i)}r_{(cm/io)}$]
f_{NGA}	= all forces acting on the vehicle that are not gravitational or aerodynamic
g_{lv}	= local gravity (plumb bob), ft/sec^2

g_m	= Newtonian mass gravitation, ft/sec^2
h	= height above the reference ellipsoid
h_{cm}	= angular momentum about mass center
H_k	= linearized measurement matrix at time t_k
$(H_k)_n$	= measurement matrix for linearized navigation equation set for NAVSTAR global positioning system (GPS LNES) solution at nth iteration
J_{cm}	= moment of inertia matrix about mass center
$(H^T H)^{-1}$	= geometry solution matrix [trace of this is geometric dilution of precision (GDOP) squared]
$N\lambda$	= number N of wavelengths λ of the L1 carrier ($N\lambda$ ambiguity)
P	= covariance matrix for random variables $(X_1, \ldots, X_n)$, $P = E(xx^T)$
$P_k(-)$	= covariance matrix before update at time t_k
$P_k(+)$	= covariance matrix after update at time t_k
Q_k	= covariance of the driving noise
R	= covariance of the measurement matrix
R_x	= correlation matrix for random variables $(X_1, \ldots, X_n)$
$R(\tau, t_1)$	= autocorrelation function
r_{A2B}	= baseline distance between users A and B
$r_{(eo/io)}$	= position vector of C_e origin "eo" with respect to C_i origin "io"
r_{eq}	= mean equatorial radius of reference ellipsoid at mean sea level
r_k	= user position relative to Earth's center: $r_k = r_{user/io} = (r_{eo/io} + \rho_k) = r_{cm/io}$
r_{mer}	= meridional radius of the ellipsoidal Earth
r_{msl}	= position vector from C_i origin to point below aircraft at mean sea level (msl)
r_{pole}	= mean polar radius of reference ellipsoid at msl
r_{prime}	= prime radius of the ellipsoidal Earth
t_k	= instant of time (discrete value)
tr	= trace of a matrix (sum of the diagonals)
u	= control vector in state-space system
u_k	= discrete control constant over interval t_k
v_k	= velocity relative to Earth-surface-fixed location = $d_e/dt[\rho_{k,v}]$, knots
$\dot{v}_{k,v}$	= $(d_v/dt)v_{k,v}$: rate of change of v_k measured by one fixed in C_v frame
v_{TAS}	= true airspeed, knots
v_{atm}	= wind velocity, knots
X	= random variable
$x_{,b}$	= vector x expressed in C_b coordinates
x_k	= discrete state-vector at time t_k
$\hat{x}_k(+)$	= error estimate after processing the update measurement
$\hat{x}_k(-)$	= error estimate before processing the update measurement
z_k	= measurement value at time t_k
$[\beta_{11} \; \beta_{12} \; \beta_{13}]$	= direction cosines of LOS vector from user's estimated position to space vehicle (SV)

$\Delta N_{\text{SV6-SV18}}(t_A)$	= unknown integer difference
$(\Delta \hat{x})_n$	= error corrections to estimate of position and time bias nth iteration
$\Delta \rho$	= difference between actual pseudorange (PR) and estimated PR
$\delta_{b/v}$	= $[\Delta \phi \ \Delta \theta \ \Delta \psi]^T$ small Euler-angle rotation vector
δ_θ	= misalignment angle vector between computer axes and true axes
δ_ϕ	= misalignment angle vector between true axes and platform axes
δ_ψ	= misalignment angle vector between computer axes and platform axes
v	= net drift due to gyro drift and torquing errors
θ	= pitch Euler angle relative to the local horizon
θ_{SV}	= angle between line of sight (LOS) to satellite SV and the horizontal
θ_p	= gyro precession angle
λ	= wavelength of L1 carrier (19 cm)
λ_c	= latitude, geocentric
λ_e	= latitude, geodetic (zero at the equator)
μ	= longitude (zero at the Greenwich Meridian)
ρ	= pseudorange between user and GPS satellite
ρ_k	= position relative to Earth surface-fixed location, nm
σ_{RMS}	= overall one sigma GPS accuracy
σ_{UERE}	= root-sum-square (RSS) of user error sources
τ	= fixed time interval
v_k	= "white" noise vector process
Φ_x	= power spectral density of a stochastic process
ϕ	= roll Euler angle
$\phi_{\text{SV6/recA}}(t_A)$	= measured carrier phase difference at time t_A by user receiver at "A" from SV6
ψ	= heading Euler angle relative to North, rad
$\Omega_{(b/i.b)}$	= matrix representation of the cross-product $\{\omega_{(b/i.b)} \times\}$ in C_b
$\omega_{b/i.b}$	= roll, pitch, and yaw rates are $[P \ Q \ R]^T$ in C_b coordinates
$\omega_{e/i}$	= angular velocity of the Earth, rad/s
ω_s	= Schuler frequency

Subscripts

eo	= origin of the C_e reference frame
eo/io	= read as eo with respect to io, as in position vector $r_{\text{eo/io}}$
,	= letter(s) following a comma in a subscript identify the reference frame

1
Navigation over Earth's Surface

This chapter begins by presenting the basic principles of navigation and position fixing necessary to understand the fundamental operation of flight navigation systems. The types of navigation systems and their properties are cataloged, and the concept of an integrated navigation system is defined. Dead reckoning navigation is discussed in detail.

The chapter continues by presenting the conventional way in which pilots locate their aircraft through the airways in the continental United States using ground-based radio navigation aids such as the VHF Omni-Directional Range (VOR) and Tactical Air Navigation (TACAN). Fundamental changes are occurring in this navigation structure because of the rapid acceptance of the Global Positioning System (GPS), a space-based radio navigational aid that is briefly introduced here and more fully described in later chapters.

A very powerful and practical way to navigate in today's airspace is to combine radio navigation aids with the position and velocity outputs of an inertial navigation system (INS). These advanced navigation applications rely on a comprehensive knowledge of the shape of the Earth and how position and velocity are obtained with respect to a reference frame fixed to the Earth's surface.

After finishing this chapter you should know the following: 1) how to represent relative position and velocity vectors with respect to the Earth; 2) the types of airborne navigation systems and their outputs that provide situational awareness to the pilot; 3) the properties of dead reckoning navigation; 4) the characteristics of and future plans for ground-based radio navigation aids; 5) the uses of radio detection and ranging (RADAR) for surveillance traffic control and precision approach; and 6) the fundamental properties of the Instrument Landing System (ILS).

1.1 Basic Principles of Navigation and Position Fixing

Navigation is the process of determining significant position, velocity, attitude, and time (PVAT) information relative to specified references. Guidance is the process of using navigation information to steer or maneuver in an intelligent, goal-seeking way. The concept of time is of the utmost importance in both processes. Navigation presumes one of the specified references is the particular location of a "user," such as the center of mass or the center of gravity of a vehicle, or the sensor location (antenna) of a black box.

For terrestrial flight the other specified reference is typically the origin of an Earth surface-fixed reference frame (commonly called a *tangent* frame). This is a Cartesian coordinate system C_e that is fixed to and rotating with the surface of the Earth with two of its axes defining the local horizontal plane. Nearly all practical navigation systems have as their goal to provide in a timely and useful manner the relative position ρ_k and relative velocity v_k of these two references. If the origin of

C_e is taken as the intersection of the Greenwich (Prime) meridian with the Earth's equator (at mean sea level), then the relative position ρ_k is commonly given as latitude, longitude, and height above mean sea level.

The Earth surface-fixed relative position ρ_k should not be confused with a vector r_k used to establish a reference location with respect to the center of the Earth. If the center of the earth is at "io" (inertia frame origin) and the center of C_e is "eo" (Earth surface-fixed origin), then $r_k = r_{\text{user/io}} = (r_{\text{eo/io}} + \rho_k)$ establishes the user position relative to Earth's center. The vector r_k defines a geocentric direction that does not exactly align with local gravity at the user's position. The misalignment occurs primarily because the Earth is not a fixed sphere, but more like a rotating ellipsoid of revolution, and because of gravity anomalies resulting from uneven mass distribution.

The concept of attitude requires the definition of two additional reference frames, both with origins at the vehicle center of mass. The first is the geographic vehicle-carried reference frame C_v with axes North (i_v), East (j_v), and Down (relative to the horizontal tangent plane at Earth's surface directly beneath the vehicle). The second is the body frame C_b with axes out the aircraft's nose (i_b), out the right wing (j_b), and down with respect to the pilot (k_b). The general orientation of these references is shown in Fig. 1.1. The relative orientation of these reference frames in flight is defined by the heading angle ψ, the pitch angle θ, and the roll angle ϕ. A pilot thinks of attitude in terms of these angles (which will be more rigorously defined later), and they are fundamental components of what pilots refer to as navigational "situational awareness." Navigation and guidance systems provide situational awareness by displaying "real time" PVAT information as needed for the mission being flown.

Note that the axes (or "basis vectors") of the vehicle-carried frame C_v are not exactly aligned with those of the Earth surface-fixed tangent frame C_e unless the

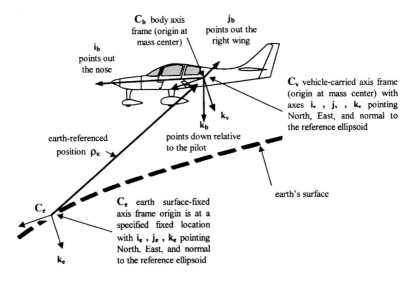

Fig. 1.1 Earth-referenced navigation.

respective origins align. Thus local "plumb bob" gravity g_{l_v} at the origin of C_v is not aligned with the local plumb bob gravity g_{l_e} at the origin of C_e (the gravity vectors are only the same if the origins align). The relative rotation rate between the C_v and C_e frames is referred to as *craft rate*. If this rate is not compensated for in some manner, an incorrect computation and display of the horizon for level flight will result in the cockpit (this error is often called *apparent precession*).

Moreover, at terrestrial locations near the poles, the orientations of the North and East axes between the two frames become especially sensitive to position. A different method of establishing a reference for navigation is thus required at high and low latitudes where the concepts of North and East are difficult to define and measure with practical accuracy.

1.1.1 Concept of Integrated Navigation Systems

Three types of navigation come to us from tradition: celestial navigation, deduced reconnaissance (commonly called dead reckoning), and pilotage. Today, modern airborne navigation systems are classified as either autonomous or position fixing, passive or active, stand-alone or aided, analog or digital. An autonomous passive system, such as an inertial guidance system, provides reasonably accurate instantaneous PVAT output with no dependence on any external man-made device or signal; that is, it is neither jammable nor capable of being sensed from outside the vehicle. Stand-alone navigation systems, on the other hand, usually require externally provided electromagnetic signals from ground-based radionavigation aids (NAVAIDS) or from space-based satellites (SV).

Integrated navigation systems combine the best features of both autonomous and stand-alone systems and are not only capable of good short-term performance in the autonomous or stand-alone mode of operation, but also provide exceptional performance over extended periods of time when in the aided mode. Integration thus brings increased performance, improved reliability and system integrity, and of course increased complexity and cost. Moreover, outputs of an integrated navigation system are usually digital, thus they are capable of being used by other resources or of being transmitted without loss or distortion.

Fundamentally, a navigation system regardless of type will output estimates of position (latitude, longitude, and height), velocity (North, East, and Down), and attitude (heading, pitch angle, roll angle). These estimates will have two types of errors associated with them: initial errors that result from the inherent accuracies of the devices and the medium through which they operate, and propagation errors that increase in time between system update measurements.

1.1.2 Dead Reckoning

Dead Reckoning (DR) computation outputs relative position by mathematically integrating relative velocity. This implies that the computer must know the initial position and the velocity estimate at a given time in vehicle carried coordinates C_v. Recall from Fig. 1.1 that this is a reference frame that is centered at the vehicle's center of mass with unit vectors i_v, j_v, k_v pointed North, East, and Down (NED). An autonomous DR system is notoriously inaccurate, and thus for aircraft navigation the system must be aided, usually with Doppler or Central Air Data Computer (CADC) relative velocity and gyroscope stabilized subsystem

(GSS) heading ψ and local vertical direction g_{l_v}. The computer must provide the magnitude of local gravity, the radius of the Earth $|r_{eo/io}| = r_{eo/io}$, time estimates, and local magnetic variation.

The computer then solves

$$\frac{d\lambda_c}{dt} = \frac{v_k \cos \psi}{r_{eo/io}} \qquad (1.1)$$

$$\frac{d\mu}{dt} = \frac{v_k \sin \psi}{r_{eo/io} \cos \lambda} \qquad (1.2)$$

where λ_c is the latitude angle and μ is the longitude angle. Note how the formulas break down at the Earth's poles where latitude is plus or minus 90°. Note also that DR can be autonomous or aided, active (Doppler) or passive, analog or digital.

DR navigation exhibits both types of navigation errors, the initial "fix" error and the propagated errors in time caused by using the preceding equations. DR systems in some cases may be integrated with radionavigation aids (NAVAIDS) to provide a continuously updated position with an autonomous capability if circumstances require it. If an autonomous DR system carried its own internal specific force receivers to determine acceleration, and its own rate sensors to determine vehicle orientation with respect to the C_e frame, it could be considered an inertial navigation system.

1.1.3 Relative Bearing and Distance

Great circles are defined relative to a spherical earth that has the same surface area as the actual earth. Great circles on the Earth have 2π radians. One degree of great circle arc is 60 n miles across the Earth's surface. One knot is a speed of 1 n mile per hour. One minute of arc on the Earth's equator defines the mean length of a nautical mile (6,076.11 ft), and a nautical mile is equivalent to 1.15 statute miles (sm). Therefore, one second of great circle arc is approximately 100 ft in length (101.3 ft), and an arc second is the angle made by Earth's radius vector $r_{eo/io}$ as its tip travels this distance along a line of constant longitude at the Earth's surface.

The zero reference for latitude and longitude is the intersection of the Greenwich meridian and the Earth's equator at mean sea level. Navigation reference points relative to this location are often called *waypoints*. Latitude (LAT) and longitude (LON) between successive waypoints are often converted into relative bearing B (degrees from local North) and relative distance d (nautical miles or degrees of great circle arc).

The conversion between nautical miles and degrees of great circle arc is complicated by the following. Although 1 n mile is always 1 min of LAT while traveling on any line of constant LON (i.e., meridian), it is not 1 min of LON while traveling on a line of constant LAT unless that LAT line is the Earth's equator. Because constant latitude lines away from Earth's equator are not great circles, on a latitude line a nautical mile is "one over the cosine of latitude" minutes of LON. Moreover, the Earth's shape is mathematically an ellipsoid. Only if relative distances between waypoints are kept short (300 n miles) can the Earth be considered a sphere, resulting in the simple dead-reckoning formulas as given here.

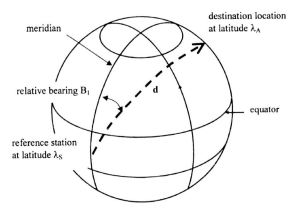

Fig. 1.2 Relative bearing and distance.

Note that the bearings between reference stations are not reciprocals of each other. This is because the respective longitude lines for the stations are not parallel. The following formulas (where all units are radians) can be used to find great circle distance d and relative bearings B_1 and B_2 from LAT and LON data. Refer to Fig. 1.2 for details.

$$\cos d = \sin \lambda_s \sin \lambda_A + \cos \lambda_s \cos \lambda_A \cos |\mu_S - \mu_A| \quad (1.3)$$

$$\sin B_1 = \frac{\cos \lambda_A \sin |\mu_S - \mu_A|}{\sin d} \quad (1.4)$$

$$\sin B_2 = \frac{\sin \lambda_A - (\sin \lambda_s \cos d)}{\cos \lambda_s \sin d} \quad (1.5)$$

The conversion between relative LAT and LON to relative bearing can be very complex for an ellipsoidal Earth (see Siouris, 1993, pp. 385–399), as can be the steering algorithms based on this information. In many of the systems described next, relative bearings and distances between the user and known Earth-referenced locations are provided electronically directly into the cockpit without requiring mathematical conversion or complex depictions of the Earth.

1.2 Air Navigation Radio Aids

Air Navigation Radio Aids are referred to here as radionavigation aids, or simply NAVAIDS. They may be ground-based or space-based, short wavelength or long wavelength. The short wavelength systems (VOR, TACAN, RADAR) are more accurate for aircraft navigation than the long wavelength systems (LORAN, OMEGA), but short wavelength systems are limited to line-of-sight (LOS) reception.

The Federal Radionavigation Plan (FRP), updated every two years to reflect government funding priorities, has put a high priority on the GPS (this system is extensively described in later sections) at the expense of the traditional ground-based NAVAIDS (some of which may be eliminated). The system description given below for the most common NAVAIDS are taken primarily from the

Aeronautical Information Manual (AIM). These descriptions are augmented where appropriate by the content of the FRP. The intent is to survey these NAVAIDS for useful flight navigation information and to recognize and anticipate the rapid change that is occurring. It should be remembered that the Federal Aviation Administration (FAA) has the sole authority to establish, operate, maintain, and prescribe standards for the operation of NAVAIDS in federally controlled airspace.

1.2.1 Nondirectional Radio Beacon

A low or medium frequency radio beacon transmits nondirectional signals whereby the pilot of an aircraft properly equipped can determine his bearing and home on the station. These facilities normally operate in the frequency band of 190 to 535 kHz and transmit a continuous carrier with either 400 or 1020 Hz modulation. All radio beacons except the compass locators transmit a continuous three-letter identification code. The radio receiver used to pick up nondirectional radio beacon (NDB) signals is the Automatic Direction Finder (ADF).

When a radio beacon is used in conjunction with the instrument landing system markers, it is called a *compass locator*. Voice transmissions are made on radio beacons unless the letter "W" (without voice) is included in the class designator. Radio beacons are subject to disturbances that may result in erroneous bearing information. Such disturbances result from such factors as lightning, precipitation static, etc. At night radio beacons are vulnerable to interference from distant stations. Nearly all disturbances that affect the ADF bearing also affect the facility's identification. Noisy identification usually occurs when the ADF needle is erratic. Voice, music, or erroneous identification may be heard when a steady false bearing is being displayed. Because ADF receivers do not have a "flag" to warn the pilot when erroneous bearing information is being displayed, the pilot should continuously monitor the NDB identification.

Bearing accuracy is 3 to 10 deg and must test within 5 deg on approaches and 10 deg elsewhere. As of 1996 there were 1575 low and medium frequency NDB transmitters, of which 728 were operated by the FAA. Little change in the status of these facilities is expected before the year 2000.

1.2.2 VHF Omni-Directional Range, VHF Omni-Directional Range/Distance Measuring Equipment, and Tactical Air Navigation

The three systems that provide the basic ground-based guidance for air navigation in the United States are VOR, distance measuring equipment (DME), and TACAN. Information provided to the aircraft pilot by VOR is the azimuth relative to the VOR ground station. DME provides a measurement of distance from the aircraft to the DME ground station. In most cases, VOR and DME are collocated as a VOR/DME facility. TACAN provides both azimuth and distance information and is used primarily by military aircraft. When TACAN is collocated with VOR, it is a VORTAC facility.

VORs are assigned frequencies in the 108 to 118 MHz frequency band, separated by 100 kHz. A VOR transmits two 30-Hz modulations resulting in a relative electrical phase angle equal to the azimuth angle of the receiving aircraft. A cardioid field pattern is produced in the horizontal plane and rotates at 30 Hz. A nondirectional (circular) 30-Hz pattern is also transmitted during the same time in all directions and is called the reference phase signal. The variable phase pattern changes phase in direct relationship to azimuth. The reference phase is frequency modulated whereas the variable phase is amplitude modulated. The receiver detects these two signals and computes the azimuth from the relative phase difference. The ground station errors are within ± 1.4 deg. The addition of course selection, receiver and flight technical errors, when combined using root-sum-squared (RSS) techniques, is calculated to be ± 4.5 deg.

VOR has line-of-sight limitations that could limit ground coverage to 30 miles or less. At altitudes above 5000 ft, the range is approximately 100 n miles, and above 20,000 ft, the range will approach 200 n miles. These stations radiate approximately 200 W. Terminal VOR stations are rated at approximately 50 W and are only intended for use within the terminal areas. VOR provides system integrity by removing a signal from use within 10 s of an out-of-tolerance condition detected by an independent monitor.

Operating on the line-of-sight principle, DME furnishes distance information with a very high degree of accuracy. Reliable signals may be received at distances up to 199 n miles at line-of-sight altitude with an accuracy of better than 1/2 mile or 3% of the distance, whichever is greater. Distance information received from DME equipment is *slant range* distance and not actual horizontal distance.

The interrogator in a DME-equipped aircraft generates a pulsed signal (interrogation) that is accepted by the transponder. In turn, the transponder generates pulsed signals (replies) that are sent back and accepted by the interrogator's tracking circuitry. Distance is then computed by measuring the total round trip time of the interrogation and its reply. The operation of DME is thus accomplished by paired pulse signals and the recognition of desired pulse spacings by the use of a decoder. The transponder must reply to all interrogators. The interrogator must measure elapsed time between interrogation and reply pulse pairs and translate this to distance. These systems are assigned in the 960 to 1213 MHz frequency band with a separation of 1 MHz. DME provides system integrity by removing a signal from use within 10 s of an out-of-tolerance condition detected by an independent monitor.

TACAN is a short-range ultra-high frequency (UHF) radionavigation signal between 960 and 1215 MHz whose transmitters and responders provide the data necessary to determine magnetic bearing and distance from an aircraft to a selected station. TACAN stations in the U.S. are frequently collocated with VOR stations. These facilities are known as VORTACS. The ground station errors are less than ± 1 deg, and distance errors are the same as DME errors. There is no ambiguity in the TACAN range information. There is a slight probability of azimuth ambiguity at multiples of 40 deg. TACAN has a line-of-sight limitation that limits ground coverage to 30 n miles or less. At altitudes of 5000 ft, the range will approach 100 n miles; above 18,000 ft, the range approaches 200 n miles. The station output power is 5 kW.

1.2.3 Instrument Landing System

Instrument Landing System (ILS) is a precision approach system normally consisting of a localizer facility, a glide slope facility, and two or three VHF marker beacons. It provides vertical and horizontal navigational (guidance) information during the approach to landing at an airport runway. At present, ILS is the primary worldwide, International Civil Aviation Organization (ICAO)-approved, precision landing system. This system is presently adequate but has limitations in siting, frequency allocation, cost, and performance.

The localizer facility and antenna are typically located 1000 ft beyond the stop end of the runway and provide a VHF (108 to 112 MHz) signal. The glide slope facility is located approximately 1000 ft from the approach end of the runway and provides a UHF (328.6 to 335.4 MHz) signal. Marker beacons are located along an extension of the runway centerline and identify particular locations on the approach. Ordinarily, two 75 MHz beacons are included as part of the instrument landing system: an outer marker at the initial approach fix (typically 4 to 7 miles from the approach end of the runway) and a middle marker located 3500 ft plus or minus 250 ft from the runway threshold. The middle marker is located so as to note impending visual acquisition of the runway in conditions of minimum visibility for Category I ILS approaches. An inner marker, located approximately 1000 ft from the threshold, is normally associated with Category II and III ILS approaches.

For a typical 10,000-ft runway, the course alignment (localizer) at threshold is maintained within ± 25 ft. Course bends during the final segment of the approach do not exceed ± 0.06 deg (2 sigma). Glide slope course alignment is maintained within ± 7 ft at 100 ft (2 sigma) elevation, and glide path bends during the final segment of the approach do not exceed ± 0.07 deg (2 sigma). See Fig. 1.3 for details.

The localizer provides course guidance throughout the descent path to the runway threshold from a distance of 18 n miles from the antenna between an altitude of 1000 ft above the highest terrain along the course line and 4500 ft above the elevation of the antenna site. Proper off-course indications are provided throughout

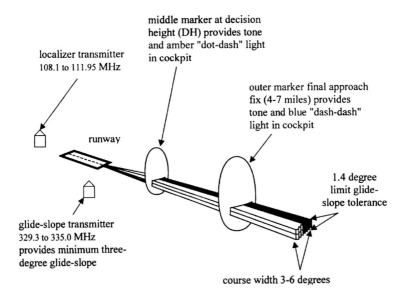

Fig. 1.3 Instrument landing system layout.

10 deg either side of the course along a radius of 18 n miles from the antenna, and from 10 to 35 deg either side of the course along a radius of 10 n miles.

The UHF glide slope transmitter, operating on one of the 40 ILS channels within the frequency range 329.15 MHz to 335.00 MHz radiates its signals in the direction of the localizer front course. False glide slope signals may exist in the area of the localizer back course approach, which can cause the glide slope flag alarm to disappear and present unreliable glide slope information.

The glide slope transmitter is located between 750 ft and 1250 ft from the approach end of the runway (down the runway) and offset 250 to 650 ft from the runway centerline. It transmits a glide path beam 1.4 deg wide. The signal provides descent information for navigation down to the lowest authorized decision height (DH) specified in the approved ILS approach procedure.

1.2.4   Long-Range Navigation and Low Frequency Radio-Navigation Aids

Long-range navigation (LORAN)-C is a pulsed, hyperbolic system, operating in the 90–110 kHz frequency band. The system is based upon measurement of the difference in time of arrival of pulses of radio frequency energy radiated by a group, or chain of transmitters, which are separated by hundreds of miles. Within a chain, one station is designated as the master station (M), and the other stations are designated as secondary stations, Whiskey (W), X-ray (X), Yankee (Y), and Zulu (Z). Signals transmitted from the secondary stations are synchronized with those from the master station. The measurement of a time-difference (TD) is made by a receiver that achieves accuracy by comparing a zero crossing of a specified radio-frequency cycle within the pulses received from the master and secondary stations of a chain. Only groundwave signals are used for air navigation in the National Airspace System (NAS). Within the groundwave range, LORAN-C will provide the user, who uses an adequate receiver, with predictable accuracy of 0.25 n mile (2 drms) or better; otherwise, predictable accuracy is 2–4 n miles. All accuracy is dependent on the user's location within the signal coverage area of the chain of stations.

Note: The use of LORAN-C for navigation in the NAS has increased considerably in recent years. In 1990, LORAN-C installations in aircraft were estimated to be in excess of 100,000. Most of these installations are for visual flight rules (VFR) use only. Approximately 10% of the aircraft LORAN-C installations are approved for instrument flight rules (IFR) use during en route and terminal operations.

OMEGA is a network of eight transmitting stations located throughout the world to provide worldwide signal coverage to Department of Defense users at a predicable accuracy of 2–4 n miles. These stations transmit in the very low frequency (VLF) band. Because of the low frequency, the signals are receivable to ranges of thousands of miles. The stations are located in Norway, Liberia, Hawaii (USA), North Dakota (USA), La Reunion, Argentina, Australia, and Japan. Accuracy is limited by the accuracy of propagation corrections transmitted to users.

Differential OMEGA stations operate on the principle of a local area monitor system comparing the received OMEGA signal with the predicted signal for the location and then transmitting a correction factor based on the observed difference. The correction factor is usually transmitted over an existing radiobeacon system and can provide an accuracy ranging from 0.3 n mile at 50 miles to 1 n mile at 500 miles.

The Federal Radionavigation Plan calls for termination of OMEGA and LORAN-C to focus on satellite navigation.

1.2.5 Radio Detection and Ranging

RADAR is a method whereby radio waves are transmitted into the air and are then received when they have been reflected by an object in the path of the beam. Range is determined by measuring the time it takes (at the speed of light) for the radio wave to go out to the object and then return to the receiving antenna. The direction of a detected object from a radar site is determined by the position of the rotating antenna when the reflected portion of the radio wave is received.

Surveillance radars are divided into two general categories: airport surveillance radar (ASR) and air route surveillance radar (ARSR). ASR is designed to provide relatively short-range coverage in the general vicinity of an airport and to serve as an expeditious means of handling terminal area traffic through observation of precise aircraft locations on a radarscope. The ASR can also be used as an instrument approach aid. ARSR is a long-range radar system designed primarily to provide a display of aircraft locations over large areas. Surveillance radars scan through 360 deg of azimuth and present target information on a radar display located in a tower or center. This information is used independently or in conjunction with other navigational aids in the control of air traffic.

Precision Approach Radar (PAR) is designed to be used as a landing aid, rather than an aid for sequencing and spacing aircraft. PAR equipment may be used as a primary landing aid, or it may be used to monitor other types of approaches. It is designed to display range, azimuth, and elevation information. Two antennas are used in the PAR array, one scanning a vertical plane, and the other scanning horizontally. Because the range is limited to 10 miles, azimuth to 20 deg, and elevation to 7 deg, only the final approach area is covered. Each scope is divided into two parts. The upper half presents altitude and distance information, and the lower half presents azimuth and distance.

1.3 Flight Management for Situational Awareness

The navigation and guidance principles and radio-navigation NAVAIDS in this chapter must provide PVAT information that is used to steer or maneuver in an intelligent, goal-seeking way, that is, to accomplish a specific mission. Often this requires combining multiple sources of horizontal guidance with multiple sources of altitude information to define routes, trajectories, or flight plans tuned to the flight characteristics of the aircraft. The user interface presented to the pilot must be intuitive, clear, and promote situational awareness.

1.3.1 Primary Flight Displays

The primary flight instrument for flight attitude orientation awareness is the attitude director indicator (ADI) or its modern equivalent, the electronic attitude director indicator (EADI). This display provides the pilot with pitch and roll angles

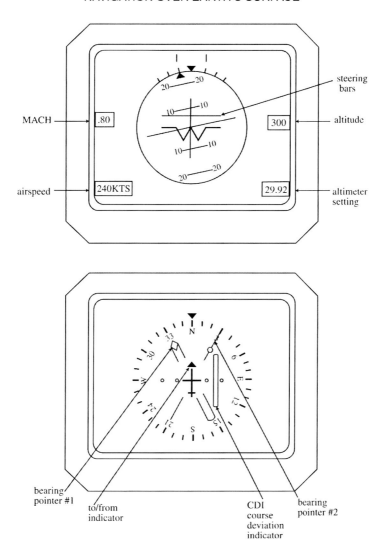

Fig. 1.4   Primary flight display.

that define the orientation of the aircraft's body axes with respect to the horizon (i.e., which way is up!). For determining navigational situational awareness, the pilot relies on the horizontal situation indicator (HSI) or its modern equivalent, the electronic horizontal situation indicator (EHSI). The combination of these two instruments, one above the other as shown in Fig. 1.4, is called the primary flight display.

Partial flight displays derived from the primary flight display are common. For example, in Fig. 1.5 a partial horizontal situation display presents a "God's eye view" of the aircraft flight path relative to selected waypoints. NAVAIDS, weather

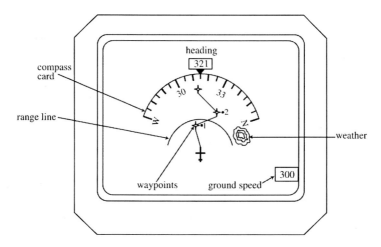

Fig. 1.5   Partial flight display.

patterns, and wind information may also be shown on this type of partial display. On modern aircraft with a glass cockpit, a selection panel with nav select buttons and bearing select knobs allows the pilots to choose the source(s) for the displayed data. The combination of these cathode-ray tube (CRT) displays, selection panels, and associated electronics is referred to as the Electronic Flight Instrumentation System (EFIS).

1.3.2   Head-Up Display

The head-up display (HUD) is an avionics module that combines optics and electronics to present key elements of the primary flight display, focused at infinity, on a combining glass located directly in line with the pilot's view out of the cockpit. This allows the aircrew to continue scanning for other aircraft, or for the runway environment during instrument conditions, without refocusing their attention back into the cockpit.

The HUD should present *only essential symbology and information specific to the task being performed*. It provides navigation and guidance information appropriate for the mode in which it operates, but nearly always shows a velocity vector, or *flight path marker*, toward which the aircraft is flying. Scales for altitude, airspeed, vertical velocity, and aircraft state are often useful, but they should be capable of being turned off at the pilot's command.

Special infrared HUD technology may enable the pilot to "see" out of the aircraft at night or during flight in instrument conditions. Low-level, high-speed flight has been demonstrated at night in mountainous terrain using forward-looking infrared (FLIR) sensors integrated with a HUD display on both F-16 and F-15 aircraft. When combined with terrain-avoidance radar, such a system is capable of autonomous operation with the pilot acting as a systems monitor. This type of HUD, however, is designed to keep the "pilot-in-the-loop" for flight control and real-time decision-making.

1.3.3 Flight Management Systems

The electronic display systems in the cockpit transfer digital data at very high rates over computer buses that conform to established standards, such as ARINC 429 for commercial aircraft and MIL-STD-1553B for military aircraft. Data from several sources are staggered in time (i.e., multiplexed) to form a pulse train on a twisted-pair conductor wire that can be routed in a relatively easy way to any place on the aircraft. Standard command-response data formats allow rapid communication between navigation, guidance, and flight control systems. The result is a large gain in flexibility, redundancy, and accuracy. The cost is complexity and possible confusion in the cockpit due to the enormous combination of selections available to the pilots.

A flight management system (FMS) is designed to eliminate cockpit confusion in the presence of electronic complexity. Position and velocity inputs are automatically scanned and blended into most probable values; flight plan computations and automatic flight control system (AFCS) signals are generated to enhance flight performance; and total overviews of the flight situation and aircraft systems are presented to the pilot on the EFIS. The FMS effectively combines the navigation computer with the CADC and thus links the black boxes hooked to navigation inputs with the black boxes hooked to the pitot static system. The term *avionics*, or aircraft electronics, is commonly used to describe the resulting complex systems.

Multifunction displays (MFDs) allow pilots to easily add, delete, or scroll through waypoints and NAVAIDS on their flight path. Pilot input to the FMS is normally accomplished with a control display unit (CDU), which is in fact a computer keyboard and display that connects the pilots to the computers that perform navigation tasks, predict flight performance, and compute guidance commands for the AFCS. These computers connect both to the NAVAIDS described in this chapter and to the inertial navigation and global positioning systems described in subsequent chapters. The computer weighting, or blending, to determine a most probable PVAT solution is accomplished with a Kalman filter, a computer algorithm that will be extensively described later in this book.

When all sensors and computer systems are working in the FMS, combined radio/inertial navigation is performed. Radio tuning is automatic given the selected waypoints and flight path. Positions and velocities are blended and displayed in the cockpit. When all the systems are not working, the FMS has flight modes for inertial only, radio only, and dead reckoning. FMS avionics are a marvel of integrated systems engineering.

Lest too much confidence be placed in these systems, remarkable as they are, it is well to remember the satirical description of the two types of pilots in the FMS glass cockpit: the enthusiastic newcomer who asks "What's it doing now?" and the seasoned veteran who replies simply "It's doing it again." Pilots always have the ultimate responsibility for situational awareness.

1.4 Closure

Lieutenant Jimmy Doolittle flew from a small field in Long Island in 1929 under a cockpit with a hood drawn over it. He made a hasty traffic pattern and landed without apparent peeking. The secrets of his success were the new flight instruments, developed by the Sperry Gyroscope Company, which provided him with crude instrumentation for heading and for attitude determination with respect to the horizon. This type of "blind" flying would give confidence to other aviation pioneers, such as Amelia Earhart and Charles Lindbergh, and to a new generation of flight engineers who would soon accurately simulate actual flight while still on the ground.

Lawrence Sperry, the founder of the company that made these instruments, invented dozens of instruments to assist flying between 1914 and 1932. One of these is still used in its original form in the modern cockpit—the turn indicator. An accomplished pilot himself, Sperry had first demonstrated it in 1915 by flying through large clouds while banking and turning, then emerging from the mists flying straight and level.

2
Newton's Laws Applied to Navigation

The basic physical law $f = ma$ only applies in a simple and direct form when in an inertial coordinate system; that is, one that is not rotating with respect to the fixed stars (such a frame is called Galilean). For practical applications the relationship is more complicated because Earth-referenced velocity must be obtained from Newton's Laws and expressed in Earth-referenced coordinates. Earth-referenced position ρ_k is then obtained by mathematically integrating relative velocity v_k (commonly called ground speed).

A firm understanding of four concepts is required to obtain v_k from Newton's Laws: gravity, attitude, Earth-referenced velocity (ground speed and direction), and the Coriolis Theorem. With these and a few basic math tools from the following chapter, a simple but practical inertial navigation block diagram can be constructed that conveys most of the important navigation concepts. This chapter starts out presenting these mathematical tools and concepts, then immediately relates them to the very important geodetic features of the Earth. A simplified version of Newton's Laws as applied to aircraft is reviewed before plunging into the Fundamental Equation of Navigation (FEN). The chapter concludes with discussions of how an inertial navigation system solves this equation and introduces the navigation satellite solution for contrast.

After finishing this chapter, you should

1) Discover how to identify the fundamental vector-matrix notation symbols used for coordinate transformations and for expressing Newton's Laws of Motion as applied to navigation.

2) Learn the concepts of gravity, attitude, Earth-referenced velocity, and the Coriolis Theorem as it relates to rotating observers.

3) Define the geodetic properties of the Earth used in flight navigation.

4) Realize the basic reference frames used in navigating over the Earth's surface, particularly the vehicle-carried vertical frame C_v and the Earth-centered, Earth-fixed frame C_{ECEF}.

5) Recognize and distinguish between the simplified aerospace vehicle equations used for aircraft motion and the more complete FEN (all of which result from $f = ma$), and appreciate the significance of the Coriolis Theorem in obtaining a solution.

6) Differentiate between the way in which a practical navigation solution is found using inertial navigation systems (accelerometers, gyros, computer, stored gravity) and the GPS (least squares triangulation plus time bias estimation); and thus understand the serious, unavoidable consquences of long-term inertial navigation caused by the inherently unstable governing equations in altitude, instrument drift, and computational approximations.

2.1 Math Tools

2.1.1 Vectors, Matrices, Reference Frames, and Direction Cosine Matrices

Vectors and matrices are represented by bold small and large letters, respectively. Coordinate systems are three-dimensional with unit basis vectors ijk (alternatively denoted 1 2 3) that are mutually perpendicular (a Cartesian dextral set). For example, C_b is the coordinate system for vehicle body-axes i_b, j_b, k_b (or $1_b, 2_b, 3_b$), which point out the nose, right-wing, and down (relative to the pilot) of an aircraft. A vector that is expressed in a particular coordinate system carries a comma notation in its subscript, followed by a letter identifying the coordinate system (the vector v_k expressed in C_v would be $v_{k,v}$).

Position vectors, such as $r_{cm/io} = r_k$ or $r_{cm/eo} = \rho_k$, should be defined with respect to a particular reference frame, but may be expressed in arbitrary coordinates, as in $r_{k,v}$ or $\rho_{k,v}$. Other vectors may be *free vectors* that do not require a specified origin in a reference frame, such as true air speed v_{TAS} or atmospheric wind velocity v_{atm}, but these also may be expressed in arbitrary coordinates. The reference frames that define a vector are thus independent of the reference frame in which the vector is expressed.

In addition to the C_b body frame, three others are important: the inertial frame C_i, the vehicle-carried vertical frame C_v, and the Earth-centered, Earth-fixed frame C_{ECEF}. An inertial reference frame is one that is not rotating with respect to the fixed stars, and it is sometimes referred to as a Galilean frame. For practical navigation, a pure Galilean C_I frame can be placed at the center of the Earth and be labeled C_i (with origin at C_{io} or io). This will result in neglecting the acceleration of the Earth around the sun, the sun about the galaxy, etc.—an accepted practice. The i_i axis is typically taken to be along the equinox and the k_i axis along the North Pole rotation axis, with j_i making the right-hand Cartesian set C_i. The C_{ECEF} frame shares the origin and k_i with C_i (that is, origin "io" = origin "ecefo" and $k_i = k_{ECEF}$) but C_{ECEF} rotates with the Earth and has its i_{ECEF} axis along the intersection of the equatorial plane and the plane containing the Geenwich Meridian.

All of this is depicted in Fig. 2.1 and may be contrasted with the original Fig. 1.1. Note that the C_v frame is a coordinate system for vehicle-carried-vertical axes (North-East-Down) with origin "vo" at the aircraft's center of mass.

A matrix as used here may be considered as a set of column vectors, each column representing either 1) the physical transformation of a basis vector, or 2) the coordinate transformation of a basis vector, or 3) both transformations simultaneously. For example, the moment of inertia matrix J relates angular velocity ω to angular momentum h, according to $h = J\omega$. Each column of J represents a physical transformation of a basis vector from angular velocity units to angular momentum units. The resulting $J\omega$ is thus a weighted sum of these transformed columns, with the weights given by the elements of ω.

A direction cosine matrix (DCM) is one whose columns exclusively represent coordinate transformations of basis vectors from one frame to another. For example, C_v^b is the direction cosine matrix, or coordinate system transformation, that re-expresses any vector originally written in C_v coordinates in C_b coordinates. This would be mathematically written as $x_{,b} = C_v^b x_{,v}$. A more useful flight relationship

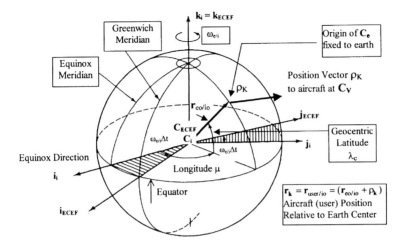

Fig. 2.1 Important reference frames for the Earth.

would be

$$v_{k.b} = C_v^b v_{k.v} = \begin{bmatrix} i_b \cdot i_v & i_b \cdot j_v & i_b \cdot k_v \\ j_b \cdot i_v & j_b \cdot j_v & j_b \cdot k_v \\ k_b \cdot i_v & k_b \cdot j_v & k_b \cdot k_v \end{bmatrix} \begin{bmatrix} v_{\text{north}} \\ v_{\text{east}} \\ v_{\text{down}} \end{bmatrix}_v \quad (2.1)$$

where the arbitrary vector is Earth-referenced velocity v_k. Note how the first column in C_v^b is the coordinate transformation of basis vector i_v into C_b coordinates, the second column expresses j_v in C_b coordinates, and the third column places k_v in C_b coordinates. This can be re-expressed as

$$v_{k.v} = C_b^v v_{k.b} = (C_b^v)(C_v^b) v_{k.v} = (C_v^b)^{-1} v_{k.b} \quad (2.2)$$

where Cartesian direction cosine matrices have the useful property that

$$(C_v^b)^{-1} = (C_v^b)^T = C_b^v = \begin{bmatrix} i_v \cdot i_b & i_v \cdot j_b & i_v \cdot k_b \\ j_v \cdot i_b & j_v \cdot j_b & j_v \cdot k_b \\ k_v \cdot i_b & k_v \cdot j_b & k_v \cdot k_b \end{bmatrix} \quad (2.3)$$

with the superscript T indicating the matrix transpose.

Note from Eqs. (2.1) and (2.3) that the DCM can be derived in a straightforward manner by projecting the basis vectors from one frame onto the other. To more easily visualize the relationships, the process may be broken down into a set of simple rotations about successive basis vectors, with each rotation resulting in a DCM. This will be illustrated in Sec. 2.3.1 for the Euler rotations ($\psi\theta\phi$) that determine the attitude of an aircraft with respect to a local-level reference frame.

Another important math tool needed to proceed is the rate of change of a vector as seen (i.e., measured) by one fixed in a rotating coordinate system. As an example, the change of Earth-referenced position ρ_k as measured by one fixed in (not rotating with respect to) the Earth-surface-fixed C_e is given by $(d_e/dt)[\rho_k]$. This is the mathematical definition of relative velocity v_k, or ground speed plus direction, one of the most important navigation system outputs:

$$\frac{d_e}{dt}r_{cm/eo} = \frac{d_e}{dt}\rho_k = v_k \tag{2.4}$$

Important Note: If a vector is expressed in the same coordinate frame that the observer is fixed in when measuring the time rate of change, total differentiation of each component is allowed; for example, using the C_e frame

$$\frac{d_e}{dt}\rho_{k.e} = \frac{d}{dt}\rho_{k.e} = \dot{\rho}_{k.e} \tag{2.5}$$

Otherwise the Coriolis Theorem, as described in a following subsection, must be applied.

The final math tools we will discuss here are the vector representation of angular velocity and the matrix representation of the cross product. They are best presented by examples. When the aircraft (assumed a rigid body) rotates with respect to inertial space, the frame C_b is rotating with respect to frame C_i. This is denoted $\omega_{b/i}$. If this angular velocity were expressed in the C_b frame, it would be written

$$\omega_{(b/i.b)} = [P \quad Q \quad R]^T \tag{2.6}$$

If this angular velocity vector were used in a cross-product operation, the symbol would be $\omega_{(b/i.b)} \times$ followed by a vector that the cross-product operates upon. This cross-product operation in navigation analysis is often replaced with its matrix equivalent. For example, if the preceding angular velocity were to operate on the aircraft's relative velocity v_k, and all quantities were expressed in C_b coordinates, then

$$\omega_{(b/i.b)} \times v_{k.b} = \Omega_{(b/i.b)} v_{k.b} \tag{2.7a}$$

where the rotation matrix associated with $\omega_{b/i.b}$

$$\Omega_{(b/i.b)} = \begin{bmatrix} 0 & -R & Q \\ R & 0 & -P \\ -Q & P & 0 \end{bmatrix} = \text{``}\omega_{(b/i.b)} \times\text{''} \tag{2.7b}$$

Note the skew-symmetric nature of the matrix representation of the cross-product with the elements of $\omega_{(b/i.b)}$ making up the elements.

2.1.2 Coriolis Theorem

This mathematical relationship, widely used in analytic mechanics, is one of the most important concepts used in navigation systems analysis and design. It states that the time rate of change of an arbitrary vector, as measured in a coordinate system C_2 rotating with respect to another coordinate system C_1, is equal to the time rate of change as measured in C_2 added to the angular velocity $\omega_{2/1}$ crossed into the arbitary vector. For instance, the time rate of change of relative position vector ρ_k as measured in an inertial coordinate system C_i, is equal to the rate of change of position vector ρ_k as measured in C_e (previously defined as relative velocity v_k) added to the angular velocity $\omega_{e/i}$ crossed into ρ_k. This can be written

NEWTON'S LAWS APPLIED TO NAVIGATION

in arbitary (that is, unspecified) coordinates as

$$\frac{d_i}{dt}\rho_k = \frac{d_e}{dt}\rho_k + (\omega_{e/i} \times \rho_k)$$
$$= v_k + \Omega_{e/i}\rho_k \tag{2.8}$$

The Coriolis Theorem will allow us to apply Newton's laws relative to inertial frame coordinates, and then to select (that is, to solve for) the relative velocity v_k that is fundamentally what we are interested in for navigation purposes. The theorem also applies to the aircraft equations derived for conventional flight mechanics (as will be shown later).

Remember not to confuse the frame that the time rate of change of a vector is "measured in" (for example, the e referring to C_e in (d_e/dt) with the frame that the vectors are "expressed in" (for example, the b referring to C_b in $v_{k,b}$). These frames are totally arbitrary and independent of one another.

2.1.3 Greenwich Mean Time

Time has never been so relative yet so accurate and universal a standard as it is today. Coordinated Universal Time (UTC) is a standard based both upon atomic clock measurements (Cesium 133 atom) and measurements of the rotation of the Earth. The atomic clock measurements form a standard called The International Atomic Time (TAI). Literally dozens of laboratories worldwide contribute to the setting of TAI, and thus it is a statistical measure that does not conform to the output of any single physical timepiece or chronometer.

The Earth rotation measurements that contribute to UTC define the precise orientation of the Earth-centered, Earth-fixed reference frame C_{ECEF} (whose i_{ECEF} axis mathematically locates the Greenwich Meridian with respect to the fixed stars). These measurements form the Universal Time 1 scales UT1 (UT1 are raw measurements UT0 corrected for polar motion) and UT2 (UT1 corrected for annual changes in Earth's rotation speed). Although scientists in Galileo's time considered the heavens an absolute time standard, modern instrumentation shows that UT1 and UT2 drift because of anomalies in Earth's rotation rate, drifts that can accumulate up to a second per year. Whenever the difference exceeds 0.9 s, UTC time "leaps" to once again march nearly in step with TAI time.

The U.S. Naval Observatory (USNO) maintains its own ensemble of atomic standards and its own time called UTC(USNO), which is kept within one microsecond of UTC. In later chapters we will see that the Global Positioning System (GPS) uses UTC(USNO) as a reference (within one microsecond), and GPS time was exactly the same value as UTC(USNO) time on January 6, 1980. Since GPS time is continuous and has no "leaps," it now (i.e., in year 1999) lags UTC(USNO) time by approximately 11 s.

One more time standard worth mentioning is ET time, for Ephemeris Time, which is obtained from the orbital motion of a planet or satellite (traditionally of the moon). The motion of the moon, incidentally, is one of the most complex gyrations observed in the heavens and was a great mystery until a great mathematician named Euler and a great German astronomer named Myer teamed up three centuries ago to solve its modes.

On test ranges a primary time standard is used called IRIG time for InterRange Instrumentation Group. IRIG time codes (A-1 kHz, B-10 kHz, and G-100 kHz for high-speed data acquisition) distribute time values and allow test events to be tagged relative to primary time standards. Range telemetry receivers have a central timing facility that receives UTC time, corrects it for any delays, and broadcasts it so that users on the test range can receive it. This subject will be covered further in the chapter on system testing.

Summarizing, *Greenwich Mean Time (GMT) is Coordinated Universal Time (UTC) and is most commonly called "Zulu" time*. It is not continuous. Time kept by the GPS master ground stations is continuous but has built up a time bias (approximately 11 s in 1999) because it never leaps to align itself with GMT.

2.2 Geodetics and Basic Reference Frames

The primary reference frames required for precise terrestrial navigation are the *body* frame C_b, the *geographic vehicle-carried* frame C_v, the *geocentric* frame C_{geo}, and the *geodetic tangent* frame C_e. Other frames of importance are the Earth-centered, Earth-fixed frame C_{ECEF} and the inertial frame C_i with origins at Earth's center of mass. Often at the start of a mission, the vehicle-carried frame C_v defines a navigation frame C_n in which the North axis i_v is rotated about k_v through wander angle α to locate i_n the local horizontal plane. Then the navigation frame i_n is not slaved to North as the vehicle travels across Earth's surface. This is important for navigation at high latitudes. These primary frames are important to mathematically define the navigation concepts of vehicle attitude, the local horizon, Newtonian and local vertical gravity, and Earth-referenced velocity.

2.2.1 Gravitation

The Newtonian acceleration of gravity due to mass attraction g_m (neglecting all stellar bodies but Earth) may be calculated for a uniform ellipsoidal Earth very accurately (within $10^{-6} g$). This is not the same as the local force of gravity g_{lv} that would be measured with a plumb bob, tape measure, and stopwatch. This is partly because the direction of g_m is only dependent on mass attraction, but the local g_{lv} includes the centripetal acceleration caused by Earth's rotation.

Based on gravity field calculations, the direction of g_m at a specified point is primarily normal to Earth's ellipsoid. Earth's centripetal acceleration will affect the meridional component of g_m in a known mathematical way. The resulting calculated local gravity field acceleration is further influenced by nonuniform mass distributions to finally produce the local gravity field g_{lv}. The angular difference between the calculated gravity field direction of Earth's homogeneous ellipsoid and the actual gravitational direction is expressed as the *deflection of the vertical*, and the difference between calculated and actual magnitudes is called the *gravity anomaly*. These concepts will be more extensively discussed in the following sections and in Chapter 3.

It is also important to understand that *gravitational acceleration must be calculated*. The human body weight seen measured on a scale, for example, is the applied specific force exerted by the body on a compressed spring and cannot be identified as inertial or gravitational without presuming the body's motion (Einstein's principle of equivalence). If the scale instantaneously accelerated with respect to the Earth, the resulting inertial acceleration applied to the human body would change the reading on the scale, but both g_m and g_{lv} would remain at their original respective values until the user position relative to Earth's mass center changed. The next section discusses g_m and g_{lv} in more detail.

Having timely access to precise values for g_m and g_{lv} on the aircraft is *one of the primary pieces of knowledge needed to run an inertial navigation system (INS)*, and the imprecision in knowing g_m and g_{lv} (plus the effects of round-off error in a computer) is one of the major limiting factors in the accuracy of our most precise inertial instrumentation. Additional knowledge about gravity, however, first requires knowledge about the Earth—its mass, shape, and dimensions. This is called *geodesy*, and it is the next topic.

2.2.2 Geodesy of the Earth

Geodesy is the branch of applied mathematics that deals with the curvature, shape, and dimensions of the Earth (*Webster's School and Office Dictionary*, 1993). Geodesy in the military has always been involved with the determination of exact positions of points on the Earth's surface for mapping or artillery control purposes, but modern requirements for precise navigation data in such fields as satellite tracking, global navigation, and missile operations have put renewed and increased emphasis on the science of geodesy.

Whereas the sphere is a close approximation of the true figure of the Earth and satisfactory for many purposes, precise measurements require that an ellipsoidal shape be fit to the Earth's surface (this was first discovered using Galileo's telescope to analyze the moons of Jupiter). Because the Earth in fact flattens slightly at the poles and bulges somewhat at the equator, the geometrical figure used in geodesy to most nearly approximate the shape of the Earth is an *ellipsoid of revolution*, obtained by rotating an ellipse about its shorter axis. It is uniquely defined by the semimajor axis and flattening. The size is represented by the radius at the equator r_{eq}, the semimajor axis, and by the radius at the poles r_{pole}, the semiminor axis. The shape of the ellipsoid is given by the flattening f, also called the "ellipticity," or by the eccentricity k_e, which are given by the relationships below

$$f = \frac{r_{eq} - r_{pole}}{r_{eq}} \qquad k_e = \left[\frac{r_{eq}^2 - r_{pole}^2}{r_{eq}^2}\right]^{\frac{1}{2}} \qquad k_e^2 = 2f\left(1 - \frac{f}{2}\right) \qquad (2.9)$$

The flattening indicates how closely an ellipsoid approaches a spherical shape. Because the flattening of the Earth is approximately 1/297, the difference between the ellipsoid of revolution representing the Earth and a sphere (the flattening for a sphere is zero) is small.

Modern technological developments have furnished new and rapid methods for data collection, and since the launching of the first Russian Sputnik, orbital data has been used to investigate and validate the ellipticity of the Earth. This has culminated in the WGS84 ellipsoid currently used by the United States. For the WGS72 USA/DoD ellipsoid of 1972, the equatorial radius was 6,378,135 m and the flattening (1/298.3).

2.2.2.1 Latitude and longitude.
The difference between geocentric latitude (λ_c = angle POA in Fig. 2.2) and geodetic latitude (λ_e = angle PDA in Fig. 2.2) is

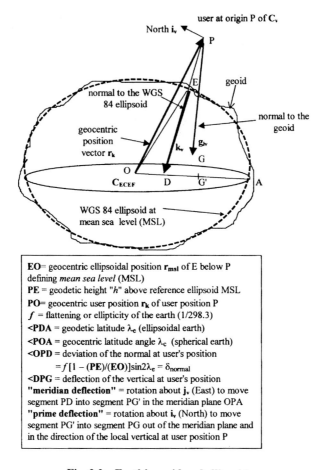

Fig. 2.2   Earth's geoid and ellipsoid.

the *deviation of the normal*. Astronomic latitude, the measurement used by ancient mariners, is the angle between local gravity and the plane of the equator. It differs from geodetic latitude at a given point on the Earth by the deflection of the vertical as previously described.

The geodetic longitude of a point is the angle between a meridian plane that contains the rotation axis of the Earth and also passes through Greenwich, England, and another meridian plane passing through the point, both planes being perpendicular to the equatorial plane. The geodetic height at a point is the distance from the reference ellipsoid at mean sea level to the point in a direction normal to the ellipsoid.

Astronomic longitude of a point on the Earth is the angle between the plane of the meridian at Greenwich (Prime Meridian) and the astronomic meridian of the point. The astronomic meridian is defined by determining the difference in time—the difference in hours, minutes, and seconds between the time a specific star is directly over the Greenwich meridian and the time the same star is directly over the meridian plane of the point. By referring to a star catalog, the exact Greenwich Mean Time the star was over the Prime Meridian is obtained. The difference between the local time at the point and the time at Greenwich is used to compute the astronomic longitude of the point. A one-hour time difference will thus become 15° of longitude difference between current position and Greenwich.

Notice how neither of the latitude measurements depends on time, but that accurate time is essential to determine longitude. The ancients could determine latitude rather easily, but the problem of determining longitude had no solution until modern times (when precise astronomical predictions were combined with precise chronometers carried on board ships to track Greenwich time).

2.2.2.2 Geoid. Figure 2.2 shows the geodetic *normal* segment PE, which by definition is perpendicular to the WGS84 ellipsoid surface. In contrast, the geoid is a surface along which the gravity potential is everywhere equal and to which the direction of local gravity is always perpendicular. This is particularly significant because optical instruments containing leveling devices are commonly used to make geodetic measurements. When properly adjusted, the vertical axis of the instrument coincides with the direction of gravity and is, therefore, perpendicular to the geoid. The geoid coincides with that surface to which the oceans would conform over the entire Earth if free to adjust to the combined effect of the Earth's mass attraction and the centrifugal force of the Earth's rotation. As a result of the uneven distribution of the Earth's mass, the geoidal surface is irregular (see Fig. 2.2).

Local gravity g_{lv} is the plumb line that is perpendicular to the geoid at the origin of C_v (sometimes called the vertical and depicted as segment PG in Fig. 2.2). The magnitude of gravity is the sum of the gravity magnitude associated with the reference ellipsoid plus a random gravity anomaly. Reference ellipsoid gravity magnitude is a complex function of gravity field parameters and latitude but is readily computed for a given position (see Siouris, 1993, page 145). The gravity anomaly is described by convention using random processes with root-mean-square (RMS) values in the range of $10^{-5}g$ to $10^{-6}g$ depending on the navigation path.

The geocentric position vector r_k is shown in Fig. 2.2 by segment OP and defines position relative to the mass center of the Earth. The perpendicular to the ellipsoid is called "the geodetic normal" and is depicted as segment PD, the meridional radius of curvature of location P (along k_v in Fig. 2.2). The angle between the two segments is called the "deviation of the normal" at the user's position. The angle between the geodetic normal and the perpendicular to the geoid (i.e., the direction of local gravity) is the "deflection of the vertical." If the meridional radius of curvature, segment PD in Fig. 2.2, were extended until it intersected the Earth's rotation axis, the resulting segment would be defined as the prime radius of curvature of the Earth's ellipsoid of location P. The two radii of curvature are used to relate the North and East components of relative velocity v_k in C_v coordinates to the rate of change of geodetic latitude λ_e and the rate of change of longitude μ.

The deviation of the normal defines the geocentric frame C_{geo} whose origin is at location P with k_{geo} along $-r_k$. The C_v frame basis vector k_v (normal to the ellipsoid) differs from C_{geo} by a rotation about $j_v = j_{geo}$ equal to δ_{normal}, the deviation of the normal, thus defining C_{geo}^v. Now assume that a common gravity model exists for the Earth, that Newtonian gravity g_m (due solely to mass field attraction) is calculated with components only along the i_{geo} and k_{geo} axes, and that the earth's mass center is its geocentric center.

Further assume that two rotations are required to move the k_v direction into g_{lv}—a very small rotation $\delta\theta_{gy}$ about j_v (East) in the meridian plane (meridian deflection) followed by a $\delta\theta_{gx}$ rotation about North (i_v for all practical purposes) in the local level plane (prime deflection). These angles are almost always smaller than one arc minute of arc regardless of location. Thus local gravity g_{lv} may be expressed as the sum of a calculated gravity field for Earth's ellipsoid plus a perturbation caused by a nonuniform mass distribution

$$g_{lv,v} = \begin{bmatrix} g\delta\theta_{gy} \\ -g\delta\theta_{gx} \\ g \end{bmatrix}_v \cong (C_{geo}^v)\{g_{m.geo} - \Omega_{e/i.geo}\Omega_{e/i.geo}r_{(cm/io.geo)}\} + \begin{bmatrix} \delta\theta_{gy} \\ -\delta\theta_{gx} \\ \Delta g \end{bmatrix}_v$$

(2.10a)

The uncertainty in knowing the small angles defining g_{lv}, combined with the anomaly in gravity magnitude, will cause errors in computed navigation outputs to propagate in time. Formulas approximating g_{lv} and g_m as a function of user distance from Earth's mass center and user latitude are commonly used in navigation calculations.

In certain types of aided-navigation simulations, the value for local gravity may be approximated by using Earth's gravitational constant G_C and a spherical Earth whose radius is the mean of the ellipsoidal polar and equatorial axes. Thus

$$g_{lv,v} \cong \begin{bmatrix} 0 \\ 0 \\ G_C/(h+r_{mean})^2 \end{bmatrix}_v \text{ m/s}^2 \qquad (2.10b)$$

where

$$G_C \cong 3.986008(10)^{14} \text{ m}^3/\text{s}^2$$

$$r_{mean} = \frac{r_{eq} + r_{pole}}{2} \cong \frac{6{,}378{,}379.79 + 6{,}362{,}576.77}{2} \text{ m} \cong 6{,}373{,}277.3 \text{ m}$$

$$h = \text{height above mean sea level (MSL) in m}$$

For an ellipsoidal Earth the preceding equation provides the familiar value of $g_{msl} \approx 9.813 \text{ m/s}^2$ for MSL ($h = 0$) at a latitude defined by r_{mean} or $g_{eq} \approx 9.780373 \text{ m/s}^2$ at the equator. More accurate approximations for gravity g_m of an ellipsoidal, nonrotating Earth include the effect of latitude directly, such as the international Airy formula for gravity at user position P (see Fig. 2.2) given by

$$g_m = g_{eq}[1 + 0.0052884 \sin^2 \lambda_c - 0.0000059 \sin^2 2\lambda_c] \qquad (2.10c)$$

where gravity for MSL at mean equator radius:

$$g_{eq} = 9.780373 \text{ m/s}^2 \qquad (2.10d)$$

Geocentric latitude at user position:

$$\lambda_c = \lambda_e - \delta_{\text{normal}} \tag{2.10e}$$

Deviation of the normal at user position:

$$\delta_{\text{normal}} = f\left(1 - \frac{h}{r_{\text{msl}}}\right) \sin 2\lambda_e \tag{2.10f}$$

Ellipsoid MSL radius at user geodetic latitude λ_e:

$$r_{\text{msl}} = r_{\text{eq}}\left(1 - f \sin^2 \lambda_e\right) \tag{2.10g}$$

If Earth centripetal acceleration is taken into account as in Eq. (2.10a), then the magnitude of local gravity may be approximated more accurately with

$$g \cong g_m - r_k \omega_{e/i}^2 \cos \lambda_e \cos \lambda_c \tag{2.10h}$$

where g_m is the gravity computed for a nonrotating Earth ellipsoid, r_k is the geocentric radius to user position $\cong (r_{\text{msl}}+h)$, and $\omega_{e/i} = 7.27220521664304(10)^{-5}$ rad/s.

Note that none of the gravity calculations in the preceding depend on longitude. The calculated value of gravity in practice is often constrained by an external source such as an altimeter to keep its value reasonable. This type of constraint and a more complete expression for all of the components of Newtonian gravity are discussed in detail in the last section of Chapter 3.

2.2.3 Navigation Outputs

Considering the impact of the preceding discussion, we can say that a modern navigation device should provide 10 outputs: geodetic latitude, longitude, and height (ρ_k); their relative change as measured by an Earth-based observer (v_k); the attitude of the aircraft (Euler angles) as defined by Fig. 2.3 and discussed in the next section; and time (either GMT time or, for satellite navigation, GPS time). For aircraft navigation there is the subsequent problem of converting geodetic height to height above the ground. A system that provides the preceding information is said to yield PVAT outputs.

2.3 Simplified Aerospace Vehicle Equations

2.3.1 Attitude of an Aircraft

Pilots are normally aware that the Euler (pronounced "oiler") angles $\phi \theta \psi$ define roll angle, pitch angle, and heading angles, respectively, which they can observe either by looking outside or by focusing on the attitude direction indicator (ADI) and the horizontal situation indicator (HSI) in the cockpit. These angles, illustrated in Fig. 2.3 can be rigorously defined mathematically by using the concepts that have been developed thus far. The Euler angles define C_v^b, the DCM that relates the vehicle-carried vertical frame to the aircraft body frame, and vice versa.

The C_v^b DCM may be derived using three successive rotations from frame C_v. The first rotation defines a new frame C_2 and is about the k_v axis until the rotated

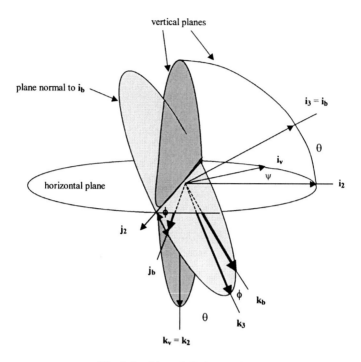

Fig. 2.3 Aircraft Euler angles.

i_v axis goes through the heading angle ψ and defines i_2. The DCM between C_v and C_2 is C_v^2. Now rotate about the j_2 (a basis vector in the local horizontal plane) so that i_2 travels through the pitch angle θ. Let this newly defined frame be C_3; the DCM between C_2 and C_3 is C_2^3. For the final third rotation, rotate about the $i_3 = i_b$ axis so that the j_3 axis rotates through roll angle ϕ and thus aligns with the direction of the right wing j_b. This final frame is C_b, and the DCM between C_3 and C_b is C_3^b. As shown in Eq. (2.1), the elements in each of these individual direction cosine matrices may be defined using the projections of basis vectors from one frame into another.

The complete DCM between the vehicle-carried vertical frame C_v and the aircraft body frame C_b may be defined using these rotations in succession as

$$C_v^b = [C_3^b(\phi)][C_2^3(\theta)][C_v^2(\psi)]$$

$$= \begin{bmatrix} 1 & 0 & 0 \\ 0 & c\phi & s\phi \\ 0 & -s\phi & c\phi \end{bmatrix} \begin{bmatrix} c\theta & 0 & -s\theta \\ 0 & 1 & 0 \\ s\theta & 0 & c\theta \end{bmatrix} \begin{bmatrix} c\psi & s\psi & 0 \\ -s\psi & c\psi & 0 \\ 0 & 0 & 1 \end{bmatrix}$$

$$= \begin{bmatrix} c\theta c\psi & c\theta s\psi & -s\theta \\ c\psi s\theta s\phi - s\psi c\phi & c\psi c\phi + s\psi s\theta s\phi & c\theta s\phi \\ c\psi s\theta c\phi + s\psi s\phi & s\psi s\theta c\phi - c\psi s\phi & c\theta c\phi \end{bmatrix} \quad (2.11)$$

NEWTON'S LAWS APPLIED TO NAVIGATION

Note that c indicates cosine and s indicates sine in the preceding matrices, and that if the elements are known, the Euler angles may be computed sequentially starting with element $(1, 3) = -\sin\theta$. It will be shown in Sec. 2.5.1 that the determination of these matrix elements is possible in real time by obtaining the angular velocity $\omega_{b/v}$ from aircraft instrumentation and computation.

To summarize the definition of Euler angles, ψ is the heading, the angle from the North to the i_b projection on the horizontal plane; θ is the pitch angle, the angle between i_b and the horizontal plane; and ϕ is the roll or bank angle, the angle j_b is rotated out of the horizontal plane about i_b. Note that heading and bank angle are undefined when pitch angle is plus or minus 90 deg.

Recall that if the Euler angles are known, then C_v^b is known and vice versa. On an aircraft the Euler angles are obtained either from specialized flight instrumentation (described in the next chapter) that is designed to output these angles, or from rate sensors (i.e., gyros) that sense $\omega_{(b/i,b)} = [P Q R]^T$ as defined by Eq. (2.6) in body coordinates. If we further define a set of Euler coordinates C_ε where i_ε is along the body axis i_b, j_ε is along j_2 (a basis vector in the local horizontal plane), and k_ε is the same as k_v that lies normal to the reference ellipsoid, then $\omega_{(b/i,b)}$ is expressed (see Fig. 2.3) in the C_b frame

$$\omega_{(b/i,b)} = C_\varepsilon^b \omega_{(b/i,\varepsilon)} = [P \quad Q \quad R]^T$$

$$\cong \begin{bmatrix} 1 & 0 & -s\theta \\ 0 & c\phi & c\theta s\phi \\ 0 & -s\phi & c\theta c\phi \end{bmatrix} \begin{bmatrix} \dot\phi \\ \dot\theta \\ \dot\psi \end{bmatrix}_\varepsilon \quad (2.12a)$$

or going the other way

$$\omega_{(b/i,\varepsilon)} = C_b^\varepsilon \omega_{(b/i,b)} \cong [\dot\phi \quad \dot\theta \quad \dot\psi]^T$$

$$\cong \begin{bmatrix} 1 & \sin\phi \tan\theta & \cos\phi \tan\theta \\ 0 & \cos\phi & -\sin\phi \\ 0 & \sin\phi \sec\theta & \cos\phi \sec\theta \end{bmatrix} \begin{bmatrix} P \\ Q \\ R \end{bmatrix}_b \quad (2.12b)$$

The sensor signals P, Q, and R ($\omega_{b/i,b}$ = the outputs of the rate gyros) and the preceding equation provide an approximation to the Euler rates ($\omega_{b/v}$), which neglects $\omega_{v/i}$. The Euler angles can then be found by integrating the Euler rates.

Note: Because the Euler rates are precisely defined as $\omega_{b/v}$, the angular velocity between frames C_b and C_v, $\omega_{v/i}$ must be taken into account for high accuracy navigation computations. The complete equation is $\omega_{b/v,\varepsilon} = C_b^\varepsilon \{\omega_{b/i,b} - C_v^b \omega_{v/i,v}\}$. The components of $\omega_{v/i,v} = \omega_{v/e,v} + \omega_{e/i,v}$ (craft rate plus Earth rate) will be derived in Sec. 2.5.1.

Note how the secant terms cause a divide-by-zero at pitch angles of plus or minus 90 deg. If the ability to determine Euler angles is lost permanently after an aircraft passes through plus or minus 90 deg of pitch, gimbal lock is said to have occurred. Recall that heading angle and roll angle are undefined at these attitudes and that gimbal lock may require a computational reset. Note also that C_ε^b is not the transpose of C_b^ε (this is because the unit basis vectors of the C_ε frame are not always orthogonal to each other—the matrix inverse must be used).

2.3.2 Quaternions

There is a practical alternative to aircraft attitude computation that eliminates the mathematical singularity at 90 deg of pitch. It is called the *quaternion*, or "four parameter" method, in which the C_b body frame is determined from the vehicle-carried geographic frame C_v using a single rotation about an axis defined solely by its direction cosines. The rotation angle and the direction cosines define the four parameters of the quaternion. Note that although the computation singularity is thus avoided, the transformation from quaternion parameters to Euler angles for display to the pilot is not possible at 90 deg of pitch. Quaternions are used extensively in the operational flight software and the simulation of highly maneuverable fighter aircraft but are not required for aircraft that seldom, if ever, approach 90 deg of pitch angle in flight.

2.3.3 Aircraft as a Rigid Body

It is very difficult to predict the general motion of an aircraft solely with mathematical equations. Fortunately, for a great variety of practical applications aircraft may be modeled as rigid bodies in which all mass particles of the vehicle have no motion relative to each other and thus rotate at the same rate about the center of mass (or equivalently about the center of gravity). A simplified version of Newton's laws may be applied to derive the conventional rigid body translational and rotational equations of motion for an aircraft. There are three translational and three rotational differential equations of motion. The three translational equations result from Newton's law of linear momentum, and the three rotational differential equations result from Newton's law of angular momentum.

The rigid body equations of motion, assuming a flat, nonrotating Earth, are normally expressed in C_b coordinates as follows:

$$F_x = \bar{q}SC_x = m(\dot{U} + QW - VR) \tag{2.13}$$

$$F_y = \bar{q}SC_y = m(\dot{V} + UR - PW) \tag{2.14}$$

$$F_z = \bar{q}SC_z = m(\dot{W} + PV - QU) \tag{2.15}$$

$$\mathscr{L} = \bar{q}S\bar{b}C_{\mathscr{L}} = \dot{P}I_x - \dot{R}I_{xz} + QR(I_z - I_y) - PQI_{xz} \tag{2.16}$$

$$\mathfrak{M} = \bar{q}S\bar{c}C_{\mathfrak{M}} = \dot{Q}I_y + PR(I_x - I_z) + (P^2 - R^2)I_{xz} \tag{2.17}$$

$$\mathfrak{N} = \bar{q}S\bar{b}C_{\mathfrak{N}} = \dot{R}I_z - \dot{P}I_{xz} + PQ(I_y - I_x) + QRI_{xz} \tag{2.18}$$

The conventional translational equations of motion are Eqs. (2.13–2.15). The conventional rotational equations of motion are the remaining three equations.

The preceding equations result from applying the Coriolis Theorem to Newton's laws. Force is mass times the time rate of change of inertial velocity with respect to an observer in inertial space. If inertial velocity can be approximated by relative Earth-referenced velocity v_k, then the time rate of change of relative velocity vector v_k, as measured in an inertial coordinate system C_i, is equal to the time rate of change of vector v_k as measured in C_b added to the angular velocity $\omega_{b/i}$ crossed into v_k. This expression of the Coriolis Theorem is given by

$$f = ma \cong m\frac{d_i}{dt}v_k$$

$$= m\left(\frac{d_b}{dt}v_k + \omega_{b/i} \times v_k\right) \quad (2.19)$$

If this vector equation is expressed in C_b coordinates, then $v_{k,b} = [U\ V\ W]^T$ and the first three force equations result [Eqs. (2.13–2.15)]. The approximation in Eq. (2.19) results because v_k is relative velocity with respect to the Earth and not with respect to C_i.

Note: To be more exact, Eq. (2.19) should use inertial velocity instead of v_k. Equations (2.25–2.29) in the next section yield a more complete answer.

For rotational motion about the center of mass, the Coriolis Theorem can be applied to the time rate of change of angular momentum $h_{cm} = J_{cm}\omega_{b/i}$ (where J_{cm} is the inertial matrix for the aircraft referenced at the center of mass). Then the time rate of change of h_{cm}, as measured in C_i, is equal to the time rate of change of vector $J_{cm}\omega_{b/i}$, as measured in C_b, added to the angular velocity $\omega_{b/i}$ crossed into $J_{cm}\omega_{b/i}$. This expression of the Coriolis Theorem, which results in the applied moment (or torque) to the aircraft about the center of mass, is given by

$$\frac{d_i}{dt}(J_{cm}\omega_{b/i}) = \frac{d_b}{dt}(J_{cm}\omega_{b/i}) + \Omega_{b/i}(J_{cm}\omega_{b/i}) \quad (2.20)$$

If this vector equation is expressed in C_b coordinates, the three moment equations result [Eqs. 2.16–2.18)].

The equations are often split into two sets that are effective in many flight situations in describing flight motions that are uncoupled into longitudinal and lateral modes. The longitudinal aircraft equations, Eqs. (2.13), (2.15), and (2.17), describe aircraft motion that is primarily confined to its own vertical plane (that is, within the plane defined by i_b and k_b). The lateral aircraft equations, Eqs. (2.14), (2.16), and (2.18), describe aircraft motion out of and away from its own vertical plane. Aircraft pitching about the j_b axis is primarily longitudinal motion, and aircraft rolling about i_b and yawing about k_b are primarily lateral motions.

2.4 Fundamental Navigation Equation

In the previous subsection Newton's laws were applied to a rigid aircraft to obtain the simplified aircraft equations of motion assuming that the Earth's surface was an inertial reference. Here the FEN will be presented, and no assumptions are made except that the inertial frame C_i may be located at the center of the Earth. The FEN is Newton's law $f = ma$; it is expressed here in C_v coordinates as

$$\frac{f_{local,v}}{m} = \dot{v}_{k,v} + \left[\Omega_{(v/i,v)} + \Omega_{(e/i,v)}\right]v_{k,v} \quad (2.21)$$

The rotation matrix $\Omega_{(e/i,v)}$ is the matrix representation of the cross-product operation $\omega_{(e/i,v)} \times$ and $v_{k,v}$ is Earth-referenced velocity (groundspeed and direction) in C_v. The local force $f_{\text{local},v}$ is related to the total force $f_{\text{total},v}$ force acting at the center of mass by

$$f_{\text{local},v} = f_{\text{total},v} - m\Omega_{e/i,v}\Omega_{e/i,v}r_{(\text{cm/io},v)}$$

$$= f_{\text{aero},v} + f_{\text{NGA},v} + (f_{G,v} + \Delta f_{G,v}) - m\Omega_{e/i,v}\Omega_{e/i,v}r_{(\text{cm/io},v)}$$

$$= f_{\text{aero},v} + f_{\text{NGA},v} + m\left\{g_{m,v} + \begin{bmatrix}\delta\theta_{gy}\\-\delta\theta_{gx}\\\Delta g\end{bmatrix}_v - \Omega_{e/i,v}\Omega_{e/i,v}r_{(\text{cm/io},v)}\right\}$$

$$= f_{\text{aero},v} + f_{\text{NGA},v} + mg_{\text{lv},v} = f_{\text{aero},v} + f_{\text{NGA},v} + f_{g,v} \qquad (2.22)$$

The total force in the preceding equations is broken down into three components: aerodynamic $f_{\text{aero},v}$ forces, not gravitational or aerodynamic $f_{\text{NGA},v}$ forces, and the local gravitational force $f_{g,v}$, which may be calculated and related to gravity $g_{\text{lv},v}$ with Eq. (2.10a). Newtonian mass attraction force $f_{G,v} = mg_{m,v}$ is calculated solely by approximating the gravity field of a nonrotating Earth ellipsoid with uniform mass distribution. An example of this calculation is presented in Sec. 3.3.6.

The remainder of this section derives Eq. (2.21). The manner in which an inertial navigation system solves this equation is described following the derivation. The reader may skip the derivation that follows and proceed to Sec. 2.5 without loss of continuity.

2.4.1 Derivation of the Fundamental Equation of Navigation

All of the math tools described in the preceding chapters will be used here and should be reviewed if necessary. The derivation in the most general case may proceed without expressing the vectors in a particular coordinate system, but C_v coordinates will be used to both fully exercise the math tools and to illustrate a commonly used local level scheme to mechanize the FEN equation for practical navigation.

A position vector expressed in C_v coordinates from the origin io of the inertial system C_i to the center of mass of the aircraft is $r_{(\text{cm/io},v)}$. The second derivative of this position vector with respect to inertial space, according to Newton, is the applied force f at the aircraft's center of mass divided by the mass. If this force is known, as well as the mass of the aircraft, then inertial acceleration may be integrated into inertial velocity and position with respect to the inertial frame at Earth's center. However, we seek relative velocity v_k and Earth-referenced position ρ_k, so the derivation should proceed in a way to yield $\dot{v}_k$, which can then be integrated into both velocity v_k and Earth-referenced position ρ_k.

Taking the first derivative of $r_{(\text{cm/io},v)}$ with respect to an inertial observer

$$\frac{d_i}{dt}r_{(\text{cm/io},v)} = v_{i,v} = \frac{d_i}{dt}\left[\rho_{k,v} + r_{(\text{eo/io},v)}\right] \qquad (2.23)$$

where v_i indicates inertial-referenced velocity. Applying Coriolis Theorem to Eq. (2.23) for an Earth-based observer in C_e so that relative velocity v_k is obtained results in

$$\frac{d_i}{dt}\left[\rho_{k.v} + r_{(eo/io.v)}\right]$$

$$= \frac{d_e}{dt}[\rho_{k.v}] + \Omega_{(e/i.v)}\rho_{k.v} + \frac{d_e}{dt}\left[r_{(eo/io.v)}\right] + \Omega_{(e/i.v)}r_{(eo/io.v)}$$

$$= \left[v_{k.v} + \Omega_{(e/i.v)}\rho_{k.v}\right] + \left[0 + \Omega_{(e/i.v)}r_{(eo/io.v)}\right]$$

$$= v_{k.v} + \Omega_{(e/i.v)}r_{(cm/io.v)} \tag{2.24}$$

When the second derivative is taken with respect to inertial space, total force f results:

$$\frac{f_{total.v}}{m} = \frac{d_i^2}{dt^2}r_{(cm/io.v)} = \frac{d_i}{dt}v_{i.v} = a_{i.v} \tag{2.25}$$

Therefore, applying Coriolis Theorem a second time to each term of Eq. (2.24) results in

$$\frac{f_{total.v}}{m} = \frac{d_v}{dt}[v_{k.v}] + \left[\Omega_{(v/i.v)}v_{k.v}\right]$$

$$+ \left[\frac{d_e}{dt}\Omega_{(e/i.v)}r_{(cm/io.v)} + \Omega_{(e/i.v)}\Omega_{(e/i.v)}r_{(cm/io.v)}\right] \tag{2.26}$$

where

$$\frac{d_e}{dt}\Omega_{(e/i.v)}r_{(cm/io.v)} = \Omega_{(e/i.v)}\frac{d_e}{dt}\rho_{k.v} = \Omega_{(e/i.v)}v_{k.v} \tag{2.27}$$

Substituting and recalling that $(d_v/dt)[v_{k.v}] = \dot{v}_{k.v}$

$$\frac{f_{total.v}}{m}$$

$$= \dot{v}_{k.v} + \left[\Omega_{(v/i.v)} + \Omega_{(e/i.v)}\right]v_{k.v} + \Omega_{(e/i.v)}\Omega_{(e/i.v)}r_{(cm/io.v)}$$

$$= \dot{v}_{k.v} + \left[\Omega_{(v/e.v)} + 2\Omega_{(e/i.v)}\right]v_{k.v} + \Omega_{(e/i.v)}\Omega_{(e/i.v)}r_{(cm/io.v)} \tag{2.28}$$

Applying Eq. (2.22) to obtain local force

$$\frac{f_{local.v}}{m} = \dot{v}_{k.v} + \left[\Omega_{(v/i.v)} + \Omega_{(e/i.v)}\right]v_{k.v} \tag{2.29}$$

which is the FEN expressed in C_v coordinates. If the derivation were accomplished using C_b coordinates, then the FEN would be written

$$\frac{f_{local.b}}{m} = \dot{v}_{k.b} + \left[\Omega_{(b/i.b)} + \Omega_{(e/i.b)}\right]v_{k.b} \tag{2.30}$$

which is left to be derived by the reader as an exercise. Note that in the preceding two equations $\dot{v}_{k.b}$ and $\dot{v}_{k.v}$ are completely different vectors (see Sec. 2.1). Thus, if the derivation were accomplished in arbitrary C_a coordinates, the FEN would be written

$$\frac{f_{\text{local}.a}}{m} = \dot{v}_{k.a} + \left[\Omega_{(a/i.a)} + \Omega_{(e/i.a)}\right]v_{k.a} \qquad (2.31)$$

or equivalently from Eq. (2.28)

$$\frac{f_{\text{local}.a}}{m} = \dot{v}_{k.a} + \left[\Omega_{(a/e.a)} + 2\Omega_{(e/i.a)}\right]v_{k.a} \qquad (2.31a)$$

where the local force in arbitrary coordinates C_a is obtained using

$$\frac{f_{\text{local}.a}}{m} = \frac{f_{\text{total}.a}}{m} - \Omega_{(e/i.a)}\Omega_{(e/i.a)} - r_{(\text{cm}/\text{io}.a)} \qquad (2.32)$$

It will be shown later that the choice of coordinate frame depends on the choice of a system mechanization to solve the FEN. This is basically a decision made about the location of the aircraft instrumentation package containing the gyros, accelerometers, and the platform on which they are mounted. For instance, if the instrument package is rigidly mounted to the airframe, then the body C_b system is chosen for the strapdown mechanization. If the instruments are isolated to float in space, then a space stable system results.

2.5 Inertial Navigation System Solution—A Preview

The computer in an inertial navigation system must solve for $\dot{v}_k$ from the FEN, regardless of the coordinates used, and integrate twice to get v_k and ρ_k, respectively. To do this, it must have an initial velocity and initial position to initiate the integration as well as continuous updates for all angular velocities and for local force f_{local}. The coordinate system chosen depends on the type of inertial navigation system, which in turn depends on the system mechanization. A strapdown INS, as stated above, has a mechanization in which the sensors are rigidly mounted to the frame of the aircraft. A gimbaled INS has a gimbaled platform that isolates the sensors from aircraft motion in a specified manner.

An INS outputs the aircraft's attitude via the Euler angles $\phi\theta\psi$. It accomplishes this by either directly reading these angles from sensors mounted on a gimbal or by computing the direction cosine matrix rate of change $\dot{C}_v^b$ (which is also a matrix) and integrating it to obtain C_v^b. The Euler angles $\phi\theta\psi$ can be computed from the elements of C_v^b using the elements of Eq. (2.11).

The outputs of an INS are capable of rapid, accurate change, and an INS is thus labeled a high bandwidth device. These outputs, moreover, are totally passive, do not depend on ground facilities or transmitted signals, and cannot be jammed or spoofed from outside the aircraft. Regardless of the precision and sensitivity of its sensors, however, the mathematical accuracy associated with computing is dependent on precise initial conditions and is susceptible to long-term drifts from the sensors as well as round-off errors within the computer. Because the unaided INS is basically a very accurate dead reckoning computer governed by equations with an unstable vertical channel, eventually it will lose its way.

The three things necessary to build an INS are specific force sensors that are used to determine local force f_{local}, rate sensors that provide rotation information

NEWTON'S LAWS APPLIED TO NAVIGATION   33

so that attitude can be determined, and a computer that stores a model of the Earth and computes local values for gravity. The INS navigation solution may be contrasted with the solution provided by the Global Positioning System provided in Sec. 2.6 at this time. The next subsection shows one way in which an INS can compute attitude information and is intended as supplementary information and as a practice exercise in applying the math tools. It may be skipped without loss of continuity.

2.5.1 Computation of the Rate of Change of Direction Cosine Matrix $\dot{C}_v^b$

As an example of how an INS computes $\dot{C}_v^b$, consider the following set of equations based on successive applications of the Coriolis Theorem:

$$\dot{v}_{k.b} = \frac{d_b}{dt} v_{k.b} = \left[\frac{d_v}{dt} v_{k.b} + \Omega_{(v/b.b)} v_{k.b} \right] \tag{2.33}$$

But note that differentiating with respect to C_v using C_b coordinates implies that

$$\frac{d_v}{dt} v_{k.b} = C_v^b \frac{d}{dt} v_{k.v} \tag{2.34}$$

and this allows Eq. (2.33) to be expressed as

$$\dot{v}_{k.b} = C_v^b \frac{d}{dt} v_{k.v} + C_v^b \Omega_{(v/b.v)} v_{k.v}$$

$$= C_v^b \frac{d}{dt} v_{k.v} + \left[C_v^b \Omega_{(v/b.v)} C_b^v \right] v_{k.b} \tag{2.35}$$

where it can be seen that the similarity transformation $[C_v^b \Omega_{(v/b.v)} C_b^v] = \Omega_{(v/b.b)}$ changes the coordinates of the rotation matrix from C_b to C_v:

$$\dot{v}_{k.b} = C_v^b \frac{d}{dt} v_{k.v} + \Omega_{(v/b.b)} v_{k.b} \tag{2.36}$$

Equation (2.33) can also be expressed as a total derivative using

$$\dot{v}_{k.b} = \frac{d_b}{dt} (C_v^b v_{k.v}) = \frac{d}{dt} (C_v^b v_{k.v}) \tag{2.37}$$

which using the Chain Rule of calculus results in

$$\dot{v}_{k.b} = \left(C_v^b \frac{d}{dt} v_{k.v} + \dot{C}_v^b v_{k.v} \right) \tag{2.38}$$

By comparing Eqs. (2.35) and (2.38) term by term, it is seen that

$$\dot{C}_v^b = C_v^b \Omega_{(v/b.v)} = -C_v^b \Omega_{(b/v.v)}$$

$$= \Omega_{(v/b.b)} C_v^b = -\Omega_{(b/v.b)} C_v^b \tag{2.39}$$

Knowing $\dot{C}_v^b$ allows $C_v^b(t_1)$ to be found from $C_v^b(t_0)$ using standard integration:

$$C_v^b(t_1) = C_v^b(t_0) + (\dot{C}_v^b)(t_1 - t_0)$$

$$= C_v^b(t_0) + C_v^b(t_0)\left[\Omega_{(v/b.v)}(t_0)\right](t_1 - t_0)$$

$$= C_v^b(t_0)\{I + \left[\Omega_{(v/b.v)}(t_0)\right](t_1 - t_0)\}$$

$$= C_v^b(t_0)\{I - \left[\Omega_{(b/v.v)}(t_0)\right](t_1 - t_0)\} \quad (2.40)$$

where $\Omega_{(v/b.v)}$ can be computed from the sensor output rotation matrix $\Omega_{(b/i.b)}$ as

$$\Omega_{(b/v.v)}(t_0) = -\Omega_{(v/b.v)}(t_0)$$

$$= \left[C_b^v(t_0)\right]\Omega_{(b/i.b)}(t_0) - \Omega_{(v/i.v)}(t_0) \quad (2.41)$$

Thus the rate sensor outputs provide $\Omega_{(b/i.b)}$, and $\Omega_{(v/i.v)}$ may be computed from

$$\Omega_{(v/i.v)} = \Omega_{(v/e.v)} + \Omega_{(e/i.v)} \quad (2.42)$$

Finally, the rotation matrix for craft rate $\Omega_{(v/e.v)}$ in Eq. (2.42) is found using Fig. 2.4 where the components of $\omega_{v/e.v}$ projected onto the C_v axes can be found to be

$$\omega_{(v/e.v)} = \begin{bmatrix} \dot{\mu}\cos\lambda_e \\ -\dot{\lambda}_e \\ -\dot{\mu}\sin\lambda_e \end{bmatrix}_v \quad (2.43)$$

and Earth rate is similarly given by

$$\omega_{(e/i.v)} = \begin{bmatrix} |\omega_{e/i}|\cos\lambda_e \\ 0 \\ -|\omega_{e/i}|\sin\lambda_e \end{bmatrix}_v \quad (2.44)$$

which, when substituted back into Eq. (2.40), yield the iterative solution for $C_v^b(t_1)$:

$$C_v^b(t_1) = C_v^b(t_0)\{I + (t_1 - t_0)\left[\Omega_{(v/e.v)}(t_0) + \Omega_{(e/i.v)}(t_0)\right]\} - (t_1 - t_0)\left[\Omega_{(b/i.b)}(t_0)\right] \quad (2.45)$$

The preceding development may be repeated to find the DCM derivative between two arbitrary coordinate frames as needed, but it has special significance when the body and vehicle-carried frames are used. For if C_v^b is known for all time (given accurate initial values and the rotation rate between the frames), then from Eq. (2.11) the Euler angles $\phi\theta\psi$ may be computed for all time and displayed in the cockpit (except of course at $\theta = \pm 90°$ when gimbal lock occurs).

Moreover, the craft and Earth rate vectors, computed from Eqs. (2.43) and (2.44) as a function of geodetic latitude and computed rates for latitude and longitude, are available to compute the axes orientation of a reference frame C_c that, in the absence of errors, will track the idealized vehicle-carried C_v axes. In this way an aircraft may store an approximation to C_v internally that is a function of computed relative velocity V_k. Thus, as seen in Fig. 2.4, relative velocity across the

NEWTON'S LAWS APPLIED TO NAVIGATION   35

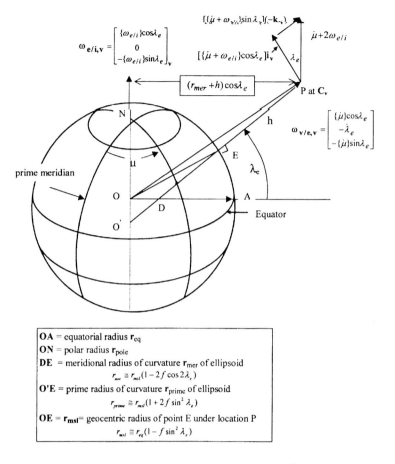

Fig. 2.4 Earth's radii and mean sea level.

reference ellipsoid determines the computed latitude and longitude rates, and these plus Earth rate provide the angular velocity $\omega_{(v/i,v)}$ to compute the axes orientation of a reference frame C_c that is intended to track by computation alone the idealized C_v frame. This will be discussed in more detail in the next few chapters.

2.6 Global Positioning System Solution—A Preview

The GPS is previewed here as a radionavigation aid for flight that provides a navigation solution for the aircraft. GPS depends on the electromagnetic spectrum as does a conventional NAVAID, but it is based on new technologies that give it enormous utility and which in many ways are refining our ideas about mapping the Earth.

The basic measurements that lead to a navigation solution using the GPS are (code) pseudo ranges often written as pseudorange. Pseudorange ρ is the magnitude

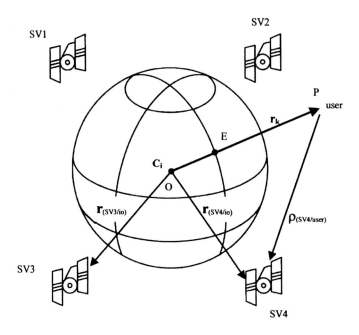

Fig. 2.5 GPS pseudorange measurements.

of $\rho_{sv/user}$, the position vector of the space vehicle satellite (sv) with respect to the user's antenna (see Fig. 2.5). A coded GPS signal is transmitted from a particular GPS satellite and is received by the GPS receiver. The GPS receiver uses correlators to measure the local transit time for the reception of a particular code phase transmitted by a GPS satellite, applies corrections, and multiplies it by the speed of light to obtain ρ. It is important not to confuse scalar pseudorange ρ with user vector position ρ_k relative to a fixed position on the Earth's surface.

After obtaining the pseudoranges to at least four different space vehicles, the GPS system computes a least-squares solution as defined by the principles of linear algebra to estimate the user position r_k relative to Earth's center. This user position is expressed in C_{ECEF} coordinates (usually in meters), which must then be converted to position ρ_k (latitude, longitude, and height). The GPS linear algebra solution also contains an estimate for the user's time correction (time bias).

2.6.1 Global Positioning System Datum Conversions

The GPS measurements must be converted from C_{ECEF} coordinates to the local datum reference that is being used for navigation. Because GPS outputs are relative to the WGS84 ellipsoid, a further conversion may be required to transform WGS84 location into local navigation coordinates. If not accomplished correctly, this can result in serious errors, especially as applied to precise ground navigation or military targeting.

The Defense Mapping Agency (DMA) was originally responsible for maintaining these conversion algorithms, but the DMA on October 1, 1996, was absorbed into a larger National Imagery and Mapping Agency (NIMA) with headquarters in Fairfax, Virginia. In recognition of its unique responsibilities and global mission, NIMA is also designated a part of the U.S. Intelligence Community, replacing the Central Imagery Office, the Defense Dissemination Program Office, the National Photographic Interpretation Center, and processing elements of the Defense Intelligence Agency, National Reconnaissance Office, Defense Airborne Reconnaissance Office, and Central Intelligence Agency. Coordinate conversions from GPS to most common map projections are available from the NIMA web site.

2.6.2 Global Positioning System Signal Path

The signal path for pseudorange $|\rho_{sv/user}|$ starts at the antenna of the space vehicle, passes through the ionosphere and troposphere where it is delayed, and is finally received at the GPS receiver's antenna. From the GPS receiver's antenna, the signal travels through a set of electronics (e.g., preamplifiers, amplifiers, frequency converters, etc.). This is a signal path that is common for all GPS satellite signals. Finally, the signal is fed into the various code correlators. These codes are pseudorandom codes and consist of a coarse code (Gold code of approximately a 1 Mhz code rate) used for acquisition called the coarse/acquisition (C/A) code and a more accurate precision (P or, when encrypted, Y) code (approximately a 10 Mhz code rate).

The GPS navigation solution requires a computer and a very specialized receiver. Relative velocity is obtained by taking the difference between successive position measurements. Standard GPS provides a PVT (position velocity time) navigation solution. There will therefore be a lag in the output display that may be significant. This would occur, for example, if GPS outputs were used in conventional course deviation displays while flying an approach. Since it takes some time for GPS outputs to be updated (typically a second or two), it is referred to as a low-bandwidth device.

Because GPS outputs do not significantly drift from correct values in time as do the outputs of an inertial navigation system, these outputs are ideally suited to aid the INS. Advanced applications, as discussed in following chapters, include in-flight alignment of an INS as well as tightly coupled systems where the GPS and INS act together continuously to provide the most probable position and velocity to a pilot.

With only one user antenna, there can be no solution for attitude using basic GPS measurements of pseudorange. GPS measurements of the carrier from multiple antennas located on a vehicle, when combined with sophisticated processing techniques, will yield attitude information (see Chapter 8). The transmitted GPS signal possesses incredible inherent accuracy and output rates that are just beginning to be exploited in applications.

To summarize, the GPS navigation solution depends on three new technologies that are relatively unique. The first technology has been discussed above and is called *pseudoranging*, or coding that accurately measures the time of arrival of a radio signal from a remote source. Next is the technology that enables precise mapping of both the Earth and its gravity field in a common world-wide reference (WGS 84 ellipsoid). Without such a common reference the conversions to other

38 INTEGRATED NAVIGATION AND GUIDANCE SYSTEMS

datums would be intractible. Finally, atomic clocks allow the precise measurement of time to greater than one part in 10 powers of 10, resulting in average navigation errors less than 100 ft.

This combination of core technologies has produced a remarkable global navigation system, but which has a low output rate for the navigation solution. Because it is drift free, it makes a suitable complement to the high-bandwidth inertial navigation solution, which, if unaided, could not compensate for its unstable altitude channel. Inertial navigation sensors and systems are discussed in the next few chapters, and GPS is treated in much more detail in Chapter 6.

2.7 Closure

In the early portion of the eighteenth century, an international race was underway to solve what was called "the longitude problem." It had always been easy to determine latitude at sea, but as another century of expansion into the New World started in 1701, no one had as yet determined how to determine longitude. A large cash prize awaited the first person to devise a scheme to determine longitude to within half a degree on the open sea. Those who attacked the problem by studying the heavens included John Flamsteed, the astronomer royal for King Charles II of England, Edmond Halley (of Halley's Comet), and Sir Isaac Newton himself.

As related by Dava Sobel in her superb book *Longitude*, Newton was impatient to test his new theory of gravitation on actual star positions being mapped by John Flamsteed, but Flamsteed was such a perfectionist that he would not release his observation records from the Royal Observatory at Greenwich Park. So Newton and his friend Halley managed to obtain the records and published a pirated copy in 1712 so that work could continue on analyzing the heavens. Flamsteed, furious at his colleagues, rounded up 300 of the 400 star catalogs that were printed and burned them. He was not willing to put his reputation on the line for work that had not yet been verified.

A modern inertial navigation system can determine the local horizontal by leveling and the direction of true North by gyrocompassing. It can even challenge the initial latitude entered by the pilot if it does not match the Earth rate component that is being measured. But to this day an inertial navigator cannot find its initial longitude without human assistance or external aiding.

3
Inertial Navigation Sensors and Systems

An inertial navigation system utilizes data from force and inertial angular velocity sensors to determine a physical body's position and velocity relative to some reference coordinate frame. Inertial navigators can be classified into three basic categories: 1) geometric, 2) semianalytic, and 3) analytic.

1) Geometric: Geometric systems were among the first inertial navigators. They provided analog information directly from the gimbal angles. To provide navigation information, as many as five gimbals may be implemented to minimize computation.

2) Semianalytic: The semianalytic system instruments only one reference frame. This frame is either an inertial nonrotating frame or a local navigation frame. At least three gimbals are needed to instrument this system, and significant computational capability is required to determine the navigation information. When the nonrotating frame is chosen as the reference frame, the system is commonly referred to as space-stabilized (SSINS). When the C_v navigation frame is chosen (local-level frame), the system is referred to as a local-level system (LLINS).

3) Analytic: A pure analytic inertial navigation system does not instrument a reference frame at all, but uses the gyro outputs to calculate the relative orientation between the system's initial and present state. This type of system is called a strapdown (SD) inertial navigation system. The lack of gimbals in the strapdown system allows for smaller designs, thus reducing size, weight, power, and sometimes cost. These systems place the maximum burden on the computational system, which in the past limited their application.

In principle, inertial navigators are well understood. On the other hand, it must be recognized that the relatively complete determination of the dynamics of an inertial navigator requires a ninth order time-varying differential equation—three degrees of freedom (DOF) in rotation about the mass center and six more DOF for position and velocity.

As with any sensor system that provides information for navigation or trajectory information, the errors of that system and their models are of primary importance. In particular it has been established that although all of the various inertial navigational systems are based upon the same physical principles, their error propagations differ depending on the chosen "mechanization" (i.e., selection of coordinate frames for computation) of the fundamental equations of navigation.

This chapter starts out with a brief description of conventional aircraft flight instruments for sensing navigation outputs, then more completely describes the specialized instrumentation used in inertial navigation systems. Basically these instruments are either force or rate sensing. The way in which these instruments are used to create each of the basic categories (platform mechanization schemes) of inertial navigation systems is then discussed, followed by an overview of system alignment techniques, including gyrocompassing. Schuler tuning is introduced as a method of bounding navigation systems errors.

The objectives of this chapter are the following:

1) Distinguish between the purpose and properties of conventional aircraft flight attitude instruments and inertial guidance instrumentation (rate and specific force sensors).

2) Describe how specific force sensors and rate sensors operate. Explain their physics, practical implications, limitations, and primary error sources (include specific force integrating receivers and ring laser gyros).

3) List the three fundamental components of an inertial navigation system (specific force sensors, rate sensors, and a computer with stored values for gravity and Earth rate) and explain how they contribute to the mechanics of an inertial platform in providing a navigation solution to the pilot. Appreciate potential dangers to flight caused by the instability of the altitude channel and why that channel must be constrained by other independent measurements.

4) Explain how the inertial navigation equations for aircraft flight are usually mechanized into either local-level or strapdown configurations and why this has important implications for both reliability and performance.

5) Understand the consequences of gimbal lock and its potential to disrupt the computer's solution to the Fundamental Equation of Navigation.

6) Describe the fundamental processes of alignment (levelling and gyrocompassing), why the pilot should accept the time penalty associated with a good alignment before flight, and realize that even with the best of alignments some unknown errors still exist that will propagate later in time.

3.1 Aircraft Gyroscopic Flight Instruments

The purpose of this subsection is to describe the basic operation of those aircraft instruments that conventionally have been used to provide outputs for attitude, attitude rates, and specific force. Many of these instruments are gyroscopic in principle and are relatively inexpensive rate and force sensors. Heading information using these basic devices is normally displayed to the aircrew on the horizontal situation indicator (HSI) using a directional gyro slaved to magnetic North. Pitch and roll attitude is displayed on the attitude director indicator (ADI) using the gyro mechanism to create *an artificial horizon*; yaw rate about the k_b axis is found on the *turn-and-slip* indicator; and specific force is indicated by needle position on the *g-meter*. Before investigating these devices, a review of the basic principles of gyroscopic action is appropriate.

3.1.1 Gyroscopic Principles

Any spinning body exhibits gyroscopic properties. A wheel designed and mounted to use these properties is called a gyroscope or gyro. Basically, a gyro is a rapidly rotating mass that is free to move about one or both axes perpendicular to the axis of rotation and to each other. The three axes of a gyro are spin axis, input (drift) axis, and output (topple) axis. In a directional gyro, the spin axis or axis of rotation is mounted horizontally. The output axis is that axis in the horizontal plane that is 90 deg from the spin axis. The input axis is that axis 90 deg vertically from the spin axis (see Fig. 3.1).

INTEGRATED NAVIGATION AND GUIDANCE SYSTEMS 41

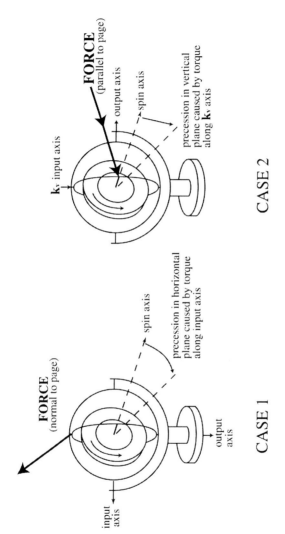

Fig. 3.1 Fundamentals of gyro precession.

A gyro is generally classified by the degrees of freedom that it possesses. A single-DOF gyro is one that is restrained so that it is free to precess about just one axis (single-DOF gyros can be further classified as rate gyros, rate integrating gyros, and integrating gyros). A two-DOF gyro is one that is free to rotate about two axes, not counting its spin axis. Two-DOF gyros are also known as displacement gyros.

Gyroscopic drift is the uncommanded rotation of the spin axis about the input (drift) axis. Topple is the rotation of the spin axis about the topple axis. These two effects result in motion of the gyro called *precession*. A freely spinning gyro tends to maintain its spin axis in a constant direction in space, a property known as rigidity in space or gyroscopic inertia. Thus, if the spin axis of a gyro were pointed toward a star, it would remain pointed in that direction in the absence of applied torque.

Apparent precession is different than precession because it is caused by the Earth rotating around a gyro fixed relative to the stars. The magnitude of apparent precession is dependent upon latitude. The horizontal component, drift, is equal to 15° per hour times the sine of the latitude. The difference in the magnitude of apparent precession caused by transporting the gyro over the Earth is called transport precession.

Real precession is caused by unwanted forces being applied to the plane of rotation of a gyro. The rotation plane then tends to rotate, not in the direction of the applied force, but 90 deg around the spin axis in the direction of spin. This is equivalent to applying a torque vector in the plane of gyro rotation. If a component of applied torque is along the input axis, the spin axis will precess toward the applied torque in the vertical plane as depicted in Fig. 3.1. If a component of torque is along the output axis, the spin axis will precess in the horizontal plane, again toward the applied torque vector.

This torquing action may be used to control the gyro by bringing about a desired reorientation of the spin axis, and most directional gyros are equipped with a device to do this. As a rule of thumb, unless restrained, *the spin axis will precess into the applied torque vector*. However, friction within the bearings of a gyro and its gimbals has the same effect as an applied torque and causes a certain amount of unwanted precession. Great care is taken in the manufacture and maintenance of gyroscopes to minimize this factor, but it is not possible to eliminate it entirely. Thus, the spin axes of all mechanical gyros will drift due to uncertainties in the applied torque unless slaved or otherwise constrained in motion.

3.1.2 Gyro-Stabilized Compass Systems

In remote-indicating gyro-stabilized compass systems, the compass direction-sensing device is outside magnetic fields created by electrical circuits in the aircraft. This is done by installing the direction-sensing device in a remote part of the aircraft such as the outer extremity of a wing. Indicators of the compass system can then be located throughout the aircraft without regard to magnetic disturbances.

Several kinds of compass systems have traditionally been used in aircraft. All include the following five basic components: 1) remote compass transmitter, 2) directional gyro, 3) amplifier, 4) heading indicators, and 5) slaving control. These compass systems are designed for airborne use at all latitudes either as a magnetic slaved compass or as a directional gyro. Many of these systems generate an electric signal that can be used as an azimuth reference by the autopilot, the radar system, the navigation and bombing computers on military aircraft, and the radio magnetic indicator.

3.1.2.1 Remote compass transmitter. This is the magnetic-direction sensing component of the compass system when the system is in operation as a magnetic-slaved compass. The transmitter is located as far from magnetic disturbances of the aircraft as possible, usually in a wing tip or the vertical stabilizer. The transmitter senses the horizontal component of the Earth's magnetic field (the only component of any value for directional purposes) and electrically transmits it to the master indicator. The compensator, an auxiliary unit of the remote compass transmitter, is used to eliminate most of the magnetic deviation caused by aircraft electrical equipment and ferrous metal, when a deviation-free location for the remote compass transmitter is not available.

3.1.2.2 Directional gyro. The directional gyro is the stabilizing component of the compass system when the system is in magnetic-slaved operation. When the compass system is in directional-gyro operation, the gyro acts as the directional reference component of the system. The directional gyro itself is restricted so that the spin axis remains parallel to the surface of the Earth. Thus, the spin axis is free to turn only in the horizontal plane (assuming the aircraft normally flies in a near-level attitude), and only the horizontal component (drift) will affect a steering gyro. In the terminology of gyro steering, precession always means the horizontal component of precession.

The actual setting of the initial reference heading is done by using the principle discussed earlier of torque application to the spinning gyro. The gyro can be set to whatever heading is desired for flight. The major error affecting the gyro and its use as a steering instrument is precession. Apparent precession will cause an apparent change of heading equal to 15° per hour times the sine of the latitude. Real precession, caused by defects in the gyro, may occur at any rate. This type of precession has been greatly reduced by the high precision of modern manufacturing methods.

3.1.3 Artificial Horizon

The use of two-DOF gyros to provide pitch and roll attitude in an aircraft's cockpit is one of the earliest applications of this technology in the air. The operation of this device depends on the gyro rotor's stability in inertial space. However, to avoid apparent precession during flight a small, pendulous mass at the bottom of the instrument is employed to slowly precess the spin axis into the direction of the *apparent vertical.* This means that the displayed artificial horizon will be *eventually* normal to the direction of the total specific force applied to the aircraft (total specific force is the total applied force on the aircraft, except for gravity, per unit mass). The word "eventually" implies that the precession may require a few minutes to occur.

If the pitch and roll angles are near zero in flight or on the ground, the direction of specific force lines up with that of local gravity and thus the artificial horizon nearly overlays the actual horizon (usually within 1/2 deg) associated with frame C_v. If the aircraft is in a steady-state turn for a prolonged period of time, the artificial horizon will precess away from the actual horizon and indicate a bank error of a few degrees after rolling out of the turn. Pilots of large, heavy aircraft that have long accelerations on take-off roll may precompensate the artificial horizon by setting a slightly incorrect pitch angle before brake release so that the indicated horizon is correct at the time of rotation.

Note: A similar idea of using a pendulous mass with a two-DOF gyroscope, but this time with the spin axis in the local horizontal plane, was used to construct a device called a mechanical gyrocompass that automatically precessed to the direction of true North. The operation of this ingenious device can be understood by imagining it located at the Earth's equator, with its spin axis in the horizontal plane but with a positive heading error of x degrees toward the East from the North Star. Without the pendulous mass, after six hours the spin axis (due to apparent precession) would be located in the local vertical plane above the North Star at x degrees above the horizon.

The existence of the pendulous mass, however, causes a torque in the horizontal plane as soon as the spin axis raises above the horizon. The spin axis then precesses into this applied torque toward the direction of true North. With appropriate mechanical damping at reasonable latitudes, and an accurate initial heading estimate, the spin axis converges on true North (plus or minus a few minutes of arc) within a few minutes of time. The use of an inertial platform for gyrocompassing will be discussed later in this chapter.

3.1.4 Turn-and-Slip Indicator

The vertical turn needle in an aircraft is a rate gyro mounted with its input axis along the k_b axis of the body frame C_b so that its output is yaw rate R. It is calibrated so that a pilot can easily accomplish a standard rate turn (3 deg of heading per second) by referring to it in flight. The yaw rate R is not necessarily the rate of change of heading angle ψ (similar to the case relating pitch rate Q and pitch angle θ). For this to occur the direction k_b (down with respect to the pilot) must be aligned with the local vertical direction k_v.

The slip indicator is a horizontal ball in the race under the turn needle, which indicates a centered (desired) position when there is no component of applied specific acceleration a_{sf} along the j_b axis of the body frame C_b (i.e., no side force or sideslip). If the ball is not centered, then the pilot will sense a sideways force. This is usually caused by the true airspeed vector making an undesired sideslip angle β on the aircraft, resulting in aerodynamics forces along the j_b axis that could upset the stability of the aircraft in extreme cases. Thus the pilot will apply a rudder input that "steps on the ball" and returns the ball to the centered position associated with coordinated flight. For slow-flying aircraft at high angle-of-attack, coordination may be difficult, especially when rapidly rolling into (or out of) turning flight. For jet aircraft at cruise speeds, coordination is not normally a problem.

3.1.5 g-Meter

The *g*-meter in the cockpit is a simple pendulous mass arranged to act as a specific force measuring device. Its sensitive axis is the k_b axis. This means that positive g forces are reaction forces that push a pilot down into the seat along the k_b axis. Some authors define positive specific force in the opposite direction; as defined here specific force is the net vector sum of all applied nongravitational forces resulting in applied specific acceleration, or a_{sf}, of the vehicle's center of mass. Applied specific acceleration must be added to local gravity g_{lv} acceleration to find the net total acceleration of the vehicle. For example, an aircraft in a steady-state, level, coordinated turn would have applied specific acceleration a_{sf} acting along the $-k_b$ axis of the aircraft, gravity acting along the k_v axis, and the vector sum of the two acting in the local horizontal plane keeping the aircraft in a circular flight path.

A g-meter, as does any accelerometer, measures all forces except gravity along its sensitive axis. When the g-meter reads one g, the aircraft is sensing the same specific force that exists when the aircraft is stationary on the ground. Zero g means that applied specific acceleration is zero and the vehicle is experiencing only local gravity g_{lv} acceleration (that is, free-fall).

3.2 Inertial Instrumentation

The flight instrumentation systems of today are more highly refined than the devices discussed above. To be suitable for use in an inertial navigation system, sensors must be classified as *precision instruments*. In this class there are two fundamental types of inertial sensors: specific force sensors and rate sensors. These sensors, and the precision platforms that they are mounted upon, are the subjects of the remainder of this chapter.

3.2.1 Specific Force Sensors

If three perfect accelerometers were mounted in an arbitrary frame C_f, each sensitive axis along a different basis vector, they would measure (output) specific force, which is the sum of all the non-gravitational forces acting on the accelerometer per unit mass. The sign of specific force is either plus or minus by convention, but to avoid confusion specific force is considered here to be applied specific acceleration a_{sf}, which is given by

$$a_{sf} = (f_{NGA} + f_{AERO})/m \qquad (3.1)$$

The total applied measurable force per unit mass in Eq. (3.1) is the sum of f_{AERO}, aerodynamic forces, and f_{NGA}, forces not gravitational or aerodynamic. Accelerometers are not capable of directly measuring gravity; thus, its value must be stored or computed as a function of position relative to the Earth.

3.2.1.1 Pendulous accelerometers. Acceleration measuring devices as shown in Fig. 3.2 are the heart of all inertial systems. It is most important that all possible sources of error be eliminated and that the accelerometers have a wide range of measurements. Changes in temperature and pressure should not affect the output of acceleration. A pendulous accelerometer consists of a pendulous mass that is free to rotate about a pivot axis in the instrument. There is an electrical pickoff that converts the rotation of the pendulous mass about its pivot axis into an output signal. This output signal is used to torque the pendulum to hold it in position, and because the signal is proportional to the measured acceleration, it is sent to the navigation computer as an acceleration output signal to be integrated for velocity and position.

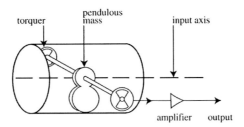

Fig. 3.2 Pendulous accelerometer.

If the torque to null the gyro output angle comes from calibrated pulses, the number of pulses may be counted and recorded in digital form (the number of these pulses per unit time is often referred to as *delta-V*). Because of the physical principles underlying its operation, this type of device has been referred to as a specific force integrating receiver (SFIR) or pulsed integrating pendulous accelerometer (PIPA).

In a pulsed integrating gyro accelerometer (PIGA) the restraining signal pulse train may be replaced by a gyro reaction caused by a pendulous float, which generates torque about the gyro output axis proportional to applied specific force along the input axis. This torque tries to precess the gyro spin axis, which is allowed the DOF to rotate about the input axis in its own internal case. A precision servo motor turns the inner case, however, at just the rate necessary to null the output of the gyro. If the outer case is space-stabilized, the relative rotation rate between the inner and outer case (obtained by counting) is an extremely accurate measure of the delta-V along the PIGA's input axis. Thus the PIGA is well known for its superb accuracy but is expensive and must be operated in a benign, stable environment.

The mathematical integration of both acceleration and velocity is very critical, and the highest accuracy is essential. There are two types of integrators, the analog and the digital. In the past one of the most common analog integrators was the resistor-capacitor (RC) amplifier, which used a charging current stabilized to a specific value proportional to an input voltage. In nearly all modern navigation systems, the integration is done numerically by digital computer.

3.2.1.2 Accelerometer errors. The pendulous accelerometer carries a liability because its sensitive axis is perpendicular to the arm of the pendulum (and lying in the plane of rotation). This means that the sensitive axis actually moves when the mass is off the null position, a factor known as sensitive axis shift. At the same time that the pendulous accelerometer becomes less sensitive to forces along its original sensitive axis (when at null), it becomes sensitive to forces at right angles to its original input axis. This condition is known as cross-coupling error.

Fortunately, the same accelerometer mechanization that minimizes nonlinearity also reduces the amount of input axis shift. The scheme is called *force rebalancing* or *torque rebalancing*. The theory behind the principle is this: take the demodulated amplified accelerometer output signal and use it to drive electromagnets; position the electromagnets to interact with permanent magnets on the accelerometer mass

INTEGRATED NAVIGATION AND GUIDANCE SYSTEMS 47

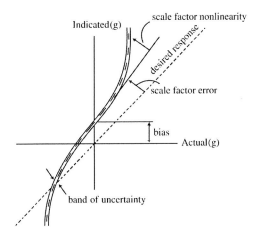

Fig. 3.3 Accelerometer errors.

to return the mass toward the null point; after the current has passed through these torque generators, drop it across a resistance to produce an analog voltage proportional to specific force.

In an aircraft, acceleration must be measured in three directions. To do this, three accelerometers should be mounted mutually perpendicular (orthogonal) in a fixed orientation, but this cannot be done perfectly. To convert acceleration into useful information, the acceleration signals must be processed and calibrated to transform the actual outputs into orthogonal axes for computation. These orientation errors are similar to the small angle attitude errors that will be discussed later in the chapter.

Each individual accelerometer is susceptible to the following (nonexclusive) list of errors, most of which are illustrated in Fig. 3.3.

Threshold. Threshold is the minimum acceleration input that causes an accelerometer electrical output. Many inertial applications require a threshold better than $10^{-6}g$.

Sensitivity. Sensitivity is the minimum change in acceleration input causing a change in accelerometer electrical output. It is usually assumed that the incremental change is made from an arbitrary, but relatively large, input compared to the incremental change.

Bias. Bias is the accelerometer electrical output under conditions of no acceleration input and is due to the residual internal forces acting on the proof mass after it has been electrically or mechanically zeroed. An INS can be designed to compensate for known accelerometer bias, provided the bias is consistent over time.

Zero uncertainty. Zero uncertainty, also referred to as null uncertainly, is the angle equivalent to the maximum variation in accelerometer output under conditions of zero acceleration input. It is caused by misalignment of the accelerometer mounting pad with the vertical datum with zero feedback current in the circuit. Errors greater than 0.1 milliradians (20 arc s) are generally not acceptable for an INS.

Zero stability. Zero stability, also referred to as null stability, is the random drift of the accelerometer output at zero acceleration input caused mainly by fluctuations in bias compensation, but could also be caused by mechanical instabilities. Typical values for an INS are of the order of ±10 arc s, or about one half of the maximum error of 20 arc s for zero position.

Scale factor errors. The proportionality constant relating the accelerometer input and output is known as the scale factor. An uncertainty or shift in the value of the scale factor causes a steady state error in the indicated acceleration. Such an error is proportional to the actual applied acceleration. A scale factor error causes an error in the accelerometer output when an acceleration is being measured (see Fig. 3.3).

Scale factor nonlinearity. In some accelerometer designs, the scale factor itself is not a constant for all ranges of applied acceleration. For example, a nonlinear spring might cause the scale factor itself to vary with acceleration. This type of error is known as scale factor nonlinearity, and if not compensated can lead to errors in indicated acceleration that are proportional to the square (or higher power) of the actual acceleration.

Time lag. The inertia of a proof mass or a pendulum prevents it from responding instantaneously to changes in acceleration. Instead, it lags such changes, thus causing temporary transient errors in the indicated acceleration. These errors are time dependent and are transient in nature, existing for a short time after any change in acceleration.

Dynamic errors. There are two dynamic errors inherent in even a perfect pendulous accelerometer: *Cross-coupling* error was discussed previously and describes errors sensitive to accelerations along an axis perpendicular to the pendulous axis. *Vibropendulosity* is a vibration induced error caused by cross coupling.

3.2.2 Precision Angular Rate Sensors

The devices described here follow the same gyroscopic principles previously described in Sec. 3.1, but are very precise. Although modern gyroscopes may have no spinning mass, they often display the same types of errors as spinning gyros. Gyros basically are characterized by their precession characteristics, as previously described, and by their rigidity (gyroscopic inertia). Rigidity is the characteristic of a gyro that makes it useful in an INS (to locate and maintain the fundamental reference frames used in navigation). The rate outputs from the gyros may be "mechanized" to produce any and all direction cosine matrices (DCMs) required for the computer to form a navigation solution. An example of how this is done for a strapdown system is presented later in this chapter.

3.2.2.1 Precision single-degree-of-freedom gyros.
The precision rate sensor depicted in Fig. 3.4 is a single-DOF gyro because it can have precession only about the output axis. It has a signal generator at one end and a torque generator at the other end. This is a hermetic integrating gyro (HIG), the type of gyro often used in missile control systems and inertial guidance systems.

3.2.2.2 Equation of motion for rate gyro.
The angle of precession about the output axis of a mechanical gyro θ_p is normally tightly constrained to be a very

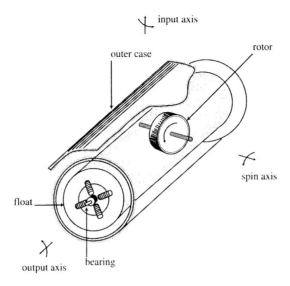

Fig. 3.4 Single degree-of-freedom gyro.

small angle. The signal required to do this, if proportional to θ_p, is thus a direct measure of the inertial angular velocity about the input axis ω_{IA}. The resulting relationship between these variables is best expressed as a transfer function in the form of a second-order damped oscillator, or

$$\left(s^2 + \frac{c_g}{I_{yf}}s + \frac{k_g}{I_{yf}}\right)\theta_p(s) = \frac{H_{\text{spin}}}{I_{yf}}\omega_{IA}(s) + U(s) \quad (3.2)$$

where I_{yf} is the effective moment of inertia about the output axis (this is greater than the physical inertia I_y due to gimbal flexure coupling into the spin angular momentum), c_g is the damping coefficient that depends on the gyro supporting fluid, k_g is the gyro elastic coefficient, and $U(s)$ represents angular acceleration uncertainties and errors. Natural frequencies near 50 Hz are common, and often such a gyro is mathematically represented simply by $Ks\theta_{IA}$ where θ_{IA} is an angle of rotation about the input axis.

This type of gyro is commonly placed on the pitch axis of an aircraft for flight control purposes so that it measures pitch rate Q about body axis j_b. Note, however, that this is not necessarily the rate of change of the Euler pitch angle θ (which is the angle between the horizontal plane and i_b). A pilot, for example, may place the aircraft into a steep bank angle near 90 deg and generate considerable pitch rate Q by applying stick back pressure with negligible change of pitch angle. For the Euler pitch angle θ to equal the pitch gyro θ_{IA}, body axis j_b must be in the horizontal plane.

3.2.2.3 Two-degree-of-freedom suspended gyro. A precision two-DOF suspended gyro is depicted in Fig. 3.5. The less accurate vertical (attitude) gyro and the directional gyro that are found in most airplanes are examples of two-

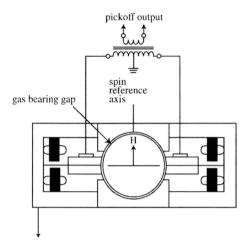

Fig. 3.5 Two degree-of-freedom suspended gyro.

DOF gyros used to track the C_v reference frame (see Sec. 3.1.3). The system as described here is used to precisely measure vehicle angular orientation for the most demanding guidance and control purposes. The bearings of an electrically suspended gyro (ESG) are nearly frictionless, and so they transmit almost no torque to the rotor. The gyro case is mounted to the structure of the vehicle, but the internal rotor orientation will remain nearly exactly fixed in space. The rotor angular momentum is made very high for rigidity by excitation with electrical eddy currents, and an optical pick-off is used to eliminate physical connections that might otherwise reduce accuracy.

3.2.2.4 Gyro errors. The errors in a gyro will depend upon its design and operating environment (i.e., single DOF, two DOF, floated in fluid, etc). Errors are normally expressed as drift rate, such as degrees per hour. Autopilot gyros found in missiles and aircraft have drift rates in the neighborhood of 50 deg per hour, whereas precision gyros used in inertial guidance systems have drifts approximating .05 deg per hour. In guidance applications the actual drift of a gyro is not nearly as important as the randomness or unpredictability of the drift. This is because there are means of compensating for a known drift by using the computer of an inertial guidance system or through changing the environment to which the gyro is subjected. Examples of such changes are heating the gyro, or orienting the gyro with respect to the acceleration vector.

The spurious nongyroscopic torques that produce unwanted drift in the gyro can be categorized as 1) constant or fixed restraint drifts, 2) mass unbalance or acceleration sensitive drifts, 3) anisoelastic or compliance drifts (drifts proportional to the square of acceleration), 4) gyro readout and torquer scale factor error, or 5) random drifts.

Nonacceleration sensitive or fixed restraint drifts are constant drifts and are generally caused by such factors as flex leads for power or signal transmission; the interaction of magnetic parts of the float with external magnetic fields (such as that of the earth); drifts caused by atmospheric changes; and drifts caused by manufacturing faults. The most common manufacturing faults take the form of sticking, which occurs when dust or small hairs get trapped between the float and the case. Other manufacturing faults include air bubbles in the flotation fluid, damaged gimbal bearings, and changes in the flotation fluid.

The gyro gimbal is never perfectly balanced about the output axis, so that any mass unbalance along the input axis or the spin axis will cause a torque about the output axis, which is proportional to acceleration. This is the predominant acceleration dependent gyro drift source (i.e., mass unbalance). Other acceleration dependent drifts result from buoyancy unbalance and changes in fluid convective currents.

Torques (drifts) that are proportional to the square of acceleration arise from non-symmetrical compliance of the gyro gimbal along the input on the spin axis under the action of steady-state acceleration or vibration. This is also called anisoelastic drift because it is caused by the elastic deformation of the gyro under acceleration.

A gyro output is usually measured in terms of voltage or current values. These electrical values must in turn be related to the physical parameter of interest (turning rate, for example, in a rate gyro). This relationship is accomplished by a scale factor K, which is a constant determined by instrument design and/or calibration. Referring to the rate gyro example, the scale factor would have units of degrees per hour per volt (or amp) output. Any error in the scale factor will result in a corresponding error in the gyro measured quantity through the auxiliary computer processing.

After all the steady-state and predictable drifts are trimmed out, there remains an irregular drift of a random nature. Included in this classification of random drifts are all predictable drift uncertainties as well as noise.

3.2.2.5 Ring laser gyros. The ring laser gyro (RLG) typically is a small, triangular device weighing a few pounds. Its principle of operation is based on the Sagnac effect and the theory of relativity (see Fig. 3.6). A resonant optical cavity hosts two traveling light waves in opposite directions. Sagnac discovered the fundamental phenomenon in 1913 that two light waves acquired a phase difference by propagating in opposite directions around a closed cavity that is rotating. Because of the precise instrumentation required and the enormous losses of the light energy,

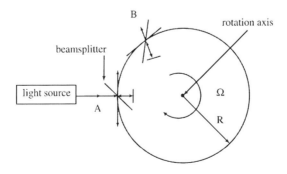

Fig. 3.6 Sagnac effect.

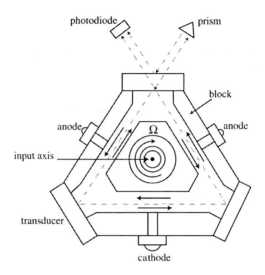

Fig. 3.7 Ring laser gyro.

the device was not practical until the laser was discovered by Macek and Davis in 1963.

The RLG is now common in nearly all inertial reference systems. Its rugged simplicity makes it ideal for strapdown INS. It is in operational use in both civil and military applications (some airliners carry three inertial platforms with RLGs). In the 1990s the Boeing Company selected the Honeywell RLG for its aircraft (Airbus selected Litton RLGs). Litton and Honeywell were involved in legal disputes of the rights to these systems for many years, especially the rights related to precision mirror technology, and so it is apparent that this is a valuable technology.

RLGs are also highly desirable angular rate sensors for use in low-cost systems. They are solid-state sensors with few moving parts that have a digital output which simplifies their interface with other avionics systems (the high rotational dynamics common in military applications are no problem for the RLG).

The RLG as illustrated in Fig. 3.7 is a unique rate integrating inertial sensor, whereby two distinct laser beams of the same frequency are emitted and split along a triangular or square path with mirrors, one beam proceeding clockwise (CW) and the other counterclockwise (CCW). Lasing is normally induced by applying a high voltage to a mixture of helium and neon gases, a process called *pumping* that results in a net amplification of photons.

If the RLG is rotating relative to inertial space, the transit times of the two beams to return to the beam-splitter will differ for each beam because the beam-splitter is also rotating, moving to a new location during the interval of time it takes the light to complete its circuit. As a result, and with respect to inertial space, the light moving in the direction of rotation must travel a longer distance than the light traveling in the opposite direction. Therefore, the two transit times will differ from the case where there is no rotation. The frequency difference between the two beams determines the output and is measured by counting pulses in the fringe pattern generated by the interference of the two light waves. The number of these pulses is proportional to the integrated angle where the difference frequency is counted as a unit of angular displacement.

The measured frequency difference is proportional to the area enclosed by the cavity of the laser and inversely proportional to the cavity length. Precision RLGs sense rotation rates of 0.001 deg per hour (0.0001 Earth rate). Frequency differences detected by the laser gyro are less than 0.1 Hz.

Laser gyros are not free of random drift. This is a major source of error that is primarily due to imperfect mirrors. There are also three commonly accepted types of errors relatively unique to a RLG: 1) null shift, 2) lock-in, or laser-lock, and 3) mode pulling. Null shift is a pulse output that occurs with no input rotation. It has a variety of causes, most of which require a knowledge of atomic physics to understand fully. Lock-in is the more troubling error for practical gyros, where backscattering and anisotropies prevent output pulses from occurring when input rotation exists. This is commonly called *deadzone*.

Dithering at a few hundred hertz is typically done to eliminate lock-in deadzone, and dithering causes a lesser statistical error called *random-walk*. A dithered RLG in a fighter aircraft is able to measure rotation rates of 400 deg per second with an accuracy of 0.01 deg per hour. Null shift affects the RLG scale factor and hinders repeatability, but does not present the practical difficulties of lock-in.

Multi-oscillator RLGs use polarization schemes implemented with quartz crystals to eliminate lock-in without using moving parts. These are low noise devices capable of rapid, accurate outputs, and they have demonstrated excellent performance in high-accuracy strapdown inertial systems.

3.2.2.6 Fiber optic gyro. Fiber-optic gyros (FOG) use optical fiber as the light path and are based on the Sagnac effect. Phase measuring electronics provide output whenever the gyro rotates with respect to inertial space. These gyros are also commonly called interferometric FOGs and consist of a diode, beamsplitter, optical fiber, and photodetector. The phase shift is proportional to the coil radius and the fiber length, as well as the input rotation rate, and so FOGs may be made very sensitive by making the optical fiber length long.

The FOG does not have lock-in and exhibits very good scale factor stability. FOGs sense rate inputs directly and do not require the standing wave that lasers provide. The FOG has no moving parts (and thus high reliability), wide dynamic range, and low cost. FOG performance, driven by commercial development of fiber optics technology, will soon meet and exceed that of the RLG. Medium accuracy gyros better than 1 deg per hour are available in 1999, and it is probable the FOG gyro will have a bright future.

3.3 Inertial Platform Mechanizations

An inertial navigator can compute ground-referenced speed and direction v_k regardless of wind and is a passive device within its operating environment. The inertial navigator provides accurate PVAT information instantaneously for maneuvers and tracking. The three basic components in any inertial navigation system are a platform oriented with rate sensors; accelerometers to supply specific components of acceleration; and a computer that integrates the signals from the sensors, models the Earth's shape plus gravity, and provides navigation output in useable formats.

Regardless of the system mechanization where the computations take place, an inertial system must eventually compute relative velocity v_k in a useful coordinate system (such as the local-level C_v frame). This normally requires that the DCM between the mechanized frame and the local-level frame be known at all times. Recall the FEN, given by Eq. (2.31), rewritten here in arbitrary mechanized C_a coordinates as

$$\dot{v}_{k.a} = \frac{f_{\text{local}.a}}{m} - [\Omega_{(a/e.a)} + 2\Omega_{(e/i.a)}]v_{k.a}$$

$$= \frac{f_{\text{local}.a}}{m} - [\Omega_{(a/i.a)} + \Omega_{(e/i.a)}]v_{k.a} \qquad \text{FEN (3.3)}$$

where the local force f_{local} is the sum of the local force of gravity f_g (local gravitational force), the aeronautical forces acting on the vehicle f_{AERO}, and all other forces f_{NGA} (where NGA indicates not gravitational or aerodynamic forces). The sum of forces at the center of mass is

$$f_{\text{local}} = f_g + f_{\text{AERO}} + f_{\text{NGA}} = m(g_{\text{lv}} + a_{\text{sf}}) \qquad (3.4)$$

where local gravity force f_g is the Newtonian gravitational force f_G (caused solely by mass attraction) modified for Earth's rotation using [see Eqs. (2.22) and (2.33)]

$$g_{\text{lv}} = \frac{f_g}{m} = \frac{f_G}{m} - \Omega_{(e/i)}\Omega_{(e/i)}r \text{ (cm/io)} \qquad (3.5)$$

The components of a_{sf} are the accelerometer outputs and measure the acceleration due to all nongravitational forces. Total acceleration caused by local force is thus seen to be

$$g_{\text{lv}} + a_{\text{sf}} = \frac{f_{\text{local}}}{m} \qquad (3.6)$$

which can be substituted into the FEN and put in C_a coordinates to yield

$$\dot{v}_{k.a} = C_v^a g_{\text{lv}.v} + C_f^a a_{\text{sf}.f} - [\Omega_{(a/e.a)} + 2\Omega_{(e/i.a)}]v_{k.a} \qquad \text{FEN (3.7)}$$

The preceding equation is mechanized in the arbitrary coordinate frame C_a. Local gravity is normally expressed in the vehicle-carried vertical frame C_v and converted into C_a coordinates. Specific acceleration, expressed in its own axis system C_f, has also been converted to C_a coordinates. If the accelerometers are mounted on a platform defining its own C_p coordinates, then $C_f^a = C_p^a C_f^p$ in the preceding equation.

Platform mechanics are associated with each different type of INS (geometric, semianalytic, and analytic). They are most commonly classified by the type of navigation coordinate frame that is implemented. The three types discussed here are the space-stable (geometric type), the local-level (semianalytic type), and the strapdown (analytic type) mechanizations.

3.3.1 Space-Stable Gimbaled Inertial Navigation System (Geometric Inertial Navigation System Type)

A free-floating platform INS is used in ballistic missiles and strategic bomber aircraft because of its great precision. These inertial navigation systems stabilize a space-stable platform using ESGs, an expensive but very accurate approach to navigation and weapons delivery.

If the arbitrary C_a and the accelerometer C_f coordinates are the C_i inertial coordinates, Eq. (3.7) may be expressed in C_i coordinates as

$$\dot{v}_{k.i} = C_v^i g_{\text{lv}.v} + a_{\text{sf}.i} - [\Omega_{(e/i.i)}] v_{k.i} \tag{3.8}$$

The relative velocity must be converted to local-level NED using the transformation C_i^v. This coordinate transformation can be found from Eq. (2.39) using

$$\dot{C}_i^v = -C_i^v \Omega_{(v/i.i)} = -\Omega_{(v/i.v)} C_i^v \tag{3.9}$$

where $\Omega_{(v/i.v)}$ is the sum of the rotation matrix for craft rate and Earth rate. The rotation vector that is needed to define this matrix was given in Eqs. (2.43) and (2.44) as

$$\omega_{v/i.v} = \omega_{v/e.v} + \omega_{e/i.v} = \begin{bmatrix} \{\dot{\mu} + \omega_{e/i}\} \cos \lambda_e \\ -\dot{\lambda}_e \\ -\{\dot{\mu} + \omega_{e/i}\} \sin \lambda_e \end{bmatrix}_v \tag{3.10}$$

$$\cong \begin{bmatrix} \left\{ \left[\dfrac{(V_{\text{EAST}})}{(r_{\text{mer}} + h) \cos \lambda_e} \right] + \omega_{e/i} \right\} \cos \lambda_e \\ -\dfrac{V_{\text{NORTH}}}{(r_{\text{mer}} + h)} \\ -\left\{ \left[\dfrac{(V_{\text{EAST}})}{(r_{\text{prime}} + h) \cos \lambda_e} \right] + \omega_{e/i} \right\} \sin \lambda_e \end{bmatrix}_v \tag{3.11}$$

where the meridional (r_{mer}) and prime (r_{prime}) radii of curvature for the ellipsoidal Earth model were defined and illustrated in Fig. 2.4. Using the flattening as defined by Eq. (2.9) and the height above the mean sea level ellipsoidal surface as h, these radii of curvature may be expressed as

Line (OE) in Fig. 2.4:

$$r_{\text{msl}} \cong r_{\text{eq}}(1 - f \sin^2 \lambda_e)$$

Flattening or ellipticity of the Earth:

$$f \cong \frac{1}{298.257}$$

Geocentric line (OP) in Fig. 2.4:

$$r_k \cong r_{\text{msl}} + h$$

Line (DE) in Fig. 2.4:

$$r_{\text{mer}} = r_{\text{msl}}(1 - 2f \cos 2\lambda_e + \text{higher order terms})$$

Line (O'E) in Fig. 2.4:

$$r_{\text{prime}} = r_{\text{msl}}\left(1 + 2f \sin^2 \lambda_e + \text{higher order terms}\right) \quad (3.12)$$

The DCM between inertial and the local-level NED frame can be initialized with geographic latitude λ_e and celestial longitude $(\mu + \omega_{e/i}t)$ using, from Fig. 2.1,

$$C_i^v = \begin{bmatrix} -\sin\lambda_e \cos(\mu + \omega_{e/i}t) & -\sin\lambda_e \sin(\mu + \omega_{e/i}t) & \cos\lambda_e \\ -\sin(\mu + \omega_{e/i}t) & \cos(\mu + \omega_{e/i}t) & 0 \\ -\cos\lambda_e \cos(\mu + \omega_{e/i}t) & -\sin(\mu + \omega_{e/i}t)\cos\lambda_e & -\sin\lambda_e \end{bmatrix} \quad (3.13)$$

An inertial mechanization that uses the C_i frame to implement the FEN has a gimbaled platform that is fixed in space, or space-stable. Errors caused by applying torque to a platform are minimized by this type of system because the core of the system, a gyro-stabilized platform called a *space integrator*, is free-floating in space. The equations for this mechanization require the least computation capability, but the system must be implemented with complex and expensive hardware.

3.3.2 Local-Level Inertial Navigation System Platforms (Semianalytic Inertial Navigation System Type)

A local-level mechanization uses the C_v frame to implement the FEN and has a platform that is torqued to keep its axes (called the C_p axes) aligned with the C_v frame. This was the most common type of practical inertial navigator for aircraft in the 1970s and 1980s. Since physical alignment of the platform was required before flight, both to align it with the computed local horizontal plane and the direction of true North, the unit required considerable warm-up time and precision in calibration.

In a local-level mechanization the accelerometers are in the C_v local-level NED frame. From Eq. (2.31) the FEN in C_v coordinates is

$$\dot{v}_{k.v} = g_{\text{lv}.v} + a_{\text{sf}.v} - \left[\Omega_{(v/i.v)} + \Omega_{(e/i.v)}\right] v_{k.v}$$

$$= g_{\text{lv}.v} + a_{\text{sf}.v} - \left[\Omega_{(v/e.v)} + 2\Omega_{(e/i.v)}\right] v_{k.v} \quad \text{FEN} \quad (3.14)$$

Although the preceding equation may be integrated directly to find $v_{k.v}$, a more common form of the FEN is obtained by differentiating, term by term,

$$v_{k.v} = \begin{bmatrix} (r_{\text{mer}} + h)\dot{\lambda}_e \\ (r_{\text{prime}} + h)(\cos\lambda_e)\dot{\mu} \\ -\dot{h} \end{bmatrix}_v \quad (3.15)$$

INTEGRATED NAVIGATION AND GUIDANCE SYSTEMS   57

and then substituting Eq. (2.10) for $g_{lv,v}$ (which includes deflection of the vertical terms), and the matrix rotation forms $\omega_{v/e,v} + 2\omega_{e/i,v}$ of Eqs. (2.43) and (2.44) so that the preceding FEN is given by

$$\begin{bmatrix} (r_{mer} + h)\ddot{\lambda}_e + (\dot{r}_{mer} + \dot{h})\dot{\lambda}_e \\ (r_{prime} + h)(\cos\lambda_e)\ddot{\mu} + (\dot{r}_{mer} + \dot{h})(\cos\lambda_e)\dot{\mu} - (\sin\lambda_e)(r_{prime} + h)\dot{\mu} \\ -\ddot{h} \end{bmatrix}_v$$

$$= \begin{bmatrix} g\delta\theta_{gy} \\ -g\delta\theta_{gx} \\ g \end{bmatrix}_v + \begin{bmatrix} a_{sfx,v} \\ a_{sfy,v} \\ a_{sfz,v} \end{bmatrix}_v$$

$$+ \begin{bmatrix} 0 & -\{\dot{\mu} + 2\omega_{e/i}\}\sin\lambda_e & \dot{\lambda}_e \\ \{\dot{\mu} + 2\omega_{e/i}\}\sin\lambda_e & 0 & \{\dot{\mu} + 2\omega_{e/i}\}\cos\lambda_e \\ -\dot{\lambda}_e & -\{\dot{\mu} + 2\omega_{e/i}\}\cos\lambda_e & 0 \end{bmatrix} v_{k,v}$$

(3.16)

The preceding coupled, nonlinear differential equation may be solved for latitude, longitude, and height above the ellipsoid by a double integration. Specific force, local gravity and its deflection angles, and initial conditions for position and velocity must be known. It is intuitive from the double integration that any constant accelerometer or gravity errors would be exponentially increasing, but it will be shown later how Schuler tuning has bounded many of these errors in the local horizontal plane. The altitude channel, however, will not have bounded error and must be constrained (see the final section of the chapter).

Note: In the development above, the local acceleration field, caused by the net effect of Newtonian gravity and centripetal acceleration, defines g_{lv}. Some authors write the preceding equation using Newtonian gravity g_m derived from gravitational potential theory. In this case Eq. (3.5) must be used to include this effect and will result in an additional term $-\Omega_{e/i,v}\Omega_{e/i,v}r_{cm/io,v}$ on the right side of Eq. (3.16). Because the geocentric position $r_k = r_{cm/io}$ does not align with the normal to the ellipsoid k_v, it is better to use $-\Omega_{e/i,v}\Omega_{e/i,v}[0, 0, (r_{prime} + h)]^T$ for the centripetal correction. The inclusion of this term in the final result is left as an exercise for the reader.

3.3.2.1 Local-level gimbals.
The platform on which the gyros and accelerometers are mounted is called a *stable element* that is isolated from the aircraft angular motions motors in C_b axes by gimbals. Torquing motors in C_p axes control the orientation of the platform as shown in Fig. 3.8. Any displacement of the stable element from its frame of reference is sensed by the electrical pick-offs in the gyroscopes. These electrical signals are amplified and used to drive the platform gimbals to realign the stable element to the local level orientation of C_v.

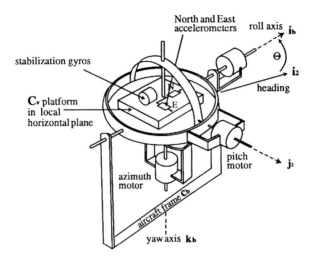

Fig. 3.8 INS stable platform.

The frame of reference for the stable element is the computed orientation of the local-level frame, or C_c. Torque motors continuously reorient C_p axes to follow C_c axes in an attempt to counter the apparent precession caused both by Earth's rotation and by v_k relative velocity. Each gyro must have its own independent operating loop. The effectiveness of the platform is determined by all parts of the platform, not just the gyros, to include torque motors, servo motors, pick-offs, amplifiers, and wiring. Gyro precession and errors caused by torquing (especially in the high latitudes) make this mechanization a less accurate but more cost effective one than the space stabilized INS.

The angles between the aircraft attitude as defined in Fig. 2.3 and the platform reference attitude are continuously measured by synchros as shown in Fig. 3.8. The aircraft yaws, rolls, and pitches about the platform in a set of gimbals, each gimbal being rotated through some component of attitude. True heading is measured between computed vectors i_2 and i_v. Roll and pitch angles are measured by synchro transmitters on the platform roll and pitch gimbals. Thus it can be seen that orienting the C_p axes is equivalent to computing an estimate of DCMs for $C_p^b C_v^p = C_v^b$.

In a conventional gimbal system the gimbals surround the platform itself. The azimuth, or innermost gimbal, carries the platform. The order of the remaining gimbals is governed by the mounting of the gimbal system in the aircraft. These gimbals are responsible for isolating the stable element from aircraft maneuvers in pitch and roll. Gimbals are named according to the function they perform and are numbered from the stable element outward: the gimbal order is normally azimuth (first), roll (second), and pitch (third). Advanced local level inertial navigation systems have a four-gimbal platform in a three-axis configuration. The fourth gimbal prevents gimbal lock at high pitch angles.

3.3.2.2 Gimbal lock. The major problem with the three gimbal system is that it is susceptible to an error called *gimbal lock*. This is possible because this gimbal system cannot isolate the stable element from all aircraft maneuvers. If the

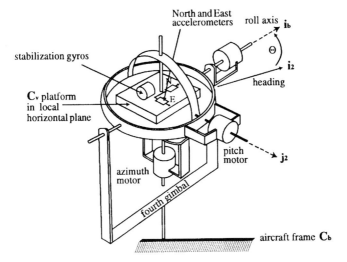

Fig. 3.9 Four-gimbal INS platform.

aircraft pitches through ±90 deg, the first gimbal axis becomes aligned with the third axis, and the stable element is no longer isolated from aircraft yaw maneuvers. This condition is known as gimbal lock and limits operations about one axis to approximately 80 deg.

A solution to the problem of gimbal lock is to use a platform that has a fourth gimbal as illustrated in Fig. 3.9. Control of this system is accomplished by the use of a pick-off between the second and third gimbal that determines the gimbal position and a torquer. In such a system, gimbal lock is prevented by driving the fourth gimbal to keep the third gimbal perpendicular to the second at all times. Any loss of orthogonality between the second and third gimbal is driven out almost instantaneously by the fourth gimbal. Therefore, the second or inner roll gimbal does not require full physical freedom for the system to operate satisfactorily. In practice the inner roll gimbal movement is limited to about ±20 deg.

3.3.2.3 Schuler tuning. Because of the double integration in Eq. (3.16) to get latitude, longitude, and height, accelerometer bias errors or tilt errors in the platform will exponentially increase and soon render the navigation solution useless. In 1923 Maxmillian Schuler noted that the motion of an imaginary pendulum, whose length would result in a period of approximately 84 min (see Fig. 3.10), would be insensitive to acceleration applied at the point of support. Although such a pendulum is obviously impractical to construct, it would indicate the vertical regardless of the acceleration of the vehicle.

Schuler compensation, as implemented in all INS systems with feedback, causes a dynamic response mode with a period of approximately 84 min that effectively prevents the unbounded propagation of certain types of system errors. It is accomplished by commanding the platform torque motors with a feedback signal based on computed relative velocity and position.

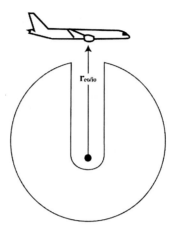

Fig. 3.10 Schuler pendulum.

Schuler compensation may be illustrated using a simple example. Assume that an INS remains located at P_A on the ellipsoidal Earth but that a misalignment θ_y angle about the east axis j_y exists when the NAV mode is engaged (see Fig. 3.11). A component of gravity $g(\sin\theta_y)$ will exist along the i_p platform axis, resulting in a computed acceleration to the North along the computed $\hat{i}_v = i_p$ axis (hats "^" as used here imply *computed* values). Because misalignments are normally small, the approximation $\sin\theta_y = \theta_y$ (radians) may be used without noticeable error.

The computed acceleration without Schuler compensation would result in an increasing relative velocity error and an exponentially increasing relative position error to the North. Schuler compensation, however, will calculate the computed latitude rate from the North component of computed $\hat{v}_k$ as

$$\dot{\hat{\lambda}}_e = \hat{V}_{\text{NORTH}}/(\hat{r}_{\text{mer}} + \hat{h}) \tag{3.17}$$

This becomes the command for the torque motor aligned along the $-j_p$ axis. As the INS incorrectly indicates motion North, the platform is rotating to eliminate computed apparent precession from the misalignment. This occurs at computed point $\hat{P}_E$ in Fig. 3.11 where platform i_p and North i_v axes align.

Thus at the computed point $\hat{P}_E$, the accelerometers no longer sense $g\theta_y$ as a horizontal acceleration, but the INS still incorrectly perceives a velocity component to the North. And so the platform will continue to be rotated until the INS computes its position at $\hat{P}_F$. From computed points $\hat{P}_E$ to $\hat{P}_F$, the Schuler compensation produces a negative misalignment angle that the INS computer will now treat as acceleration to the South. Finally, at $\hat{P}_F$ the North velocity is computed zero, but the perceived acceleration to the South is a maximum. The platform i_p axis will thus be torqued with Schuler compensation to rock back and forth tracking the computed $\hat{i}_v$ axis. The velocity errors in such a scheme will be bounded despite the double integration between acceleration and position error.

INTEGRATED NAVIGATION AND GUIDANCE SYSTEMS   61

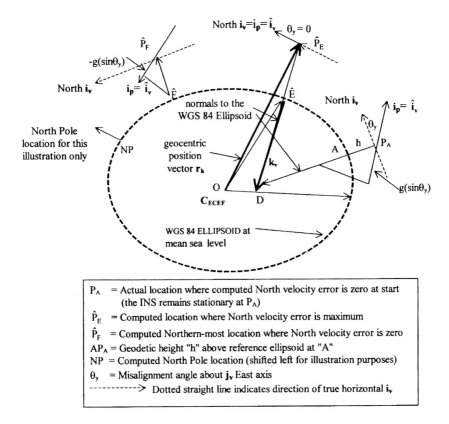

Fig. 3.11   Schuler-computed compensation for a stationary INS with initial tilt error.

Figure 3.12 is a visual picture of what is happening at the platform following the initial misalignment error. Note the Schuler compensation in this case results in a rocking platform that has bounded the velocity and position errors computed by the INS. This is more readily discerned from the block diagram of Fig. 3.13, where the transfer function for $\theta_y(s)$ from the block diagram clearly shows a cosine function in time resulting from the misalignment angle. Thus the double integration of acceleration error into position error, which would make any inertial system useless after a short time, is effectively constrained into an error that oscillates slowly from plus to minus.

Note from the block diagram of Fig. 3.13 that the response to a constant East gyro drift rate (assuming the actual position is stationary on the Earth) is given by

$$\hat{V}_{\text{NORTH}}(s) = \frac{\text{Drift Rate}}{s} \frac{\hat{g}}{\{s^2 + [\hat{g}/(\hat{r}_{\text{mer}} + \hat{h})]\}} \quad (3.18)$$

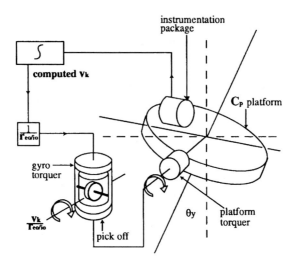

Fig. 3.12 Platform rocking caused by Schuler compensation.

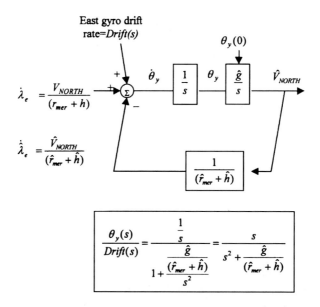

Fig. 3.13 Schuler compensation block diagram.

The inverse transform of the preceding equation provides the North velocity error response in time:

$$\hat{V}_{\text{NORTH}}(t) = (\text{Constant})\{1 - \cos \omega_s t\} \quad (3.19)$$

where

$$\omega_s = \sqrt{\frac{\hat{g}}{(\hat{r}_{\text{mer}} + \hat{h})}} \cong \sqrt{\frac{g}{|\mathbf{r}_{\text{eo/io}}|}} \cong 0.00124 \text{ rad/s} \quad (3.20)$$

$$T_{\text{Schuler}} \cong 2\pi \sqrt{\left|\frac{\mathbf{r}_{\text{eo/io}}}{g_v}\right|} \cong 84.4 \text{ min} \quad (3.21)$$

which shows a bounded North velocity error, but whose integral is an unbounded East position error that linearly increases in time plus an oscillation at ω_s, the Schuler frequency.

The net result of the Schuler feedback scheme is to torque the platform by calculating estimates for the apparent precession rates about each axis in the C_v frame so that the resulting forced precession can counter it. Even in other mechanizations where the Schuler compensation is more difficult to visualize, the Schuler oscillation will be present because of the need to calculate the estimated orientation of the local-level C_v frame for practical navigation across the surface of the Earth.

3.3.2.4 Local-level Inertial Navigation System restrictions.

Restrictions using the local vertical North-pointing INS system are encountered at the higher latitudes. These restrictions are summarized as follows:

1) The torquing rate to the azimuth gyro to compensate for transport rate becomes excessively high as latitude approaches $\pm 90°$. Depending on the velocity of the vehicle, the limiting latitude is approximately $\pm 75°$.

2) The calculation of longitude also gets difficult, and the limiting latitude is normally $\pm 80°$ to $\pm 85°$.

3) The magnitude of the error signal that is used to torque the azimuth gyro during the gyro compassing phase of alignment becomes very small as latitude increases. For practical purposes this limits gyrocompassing to latitudes of less than $\pm 75°$ to $\pm 80°$.

Some inertial systems use a principle known as wander angle, which does not require the gyros to be oriented to true North. A wander angle inertial system has the advantage of being able to operate in polar regions where excessive torquing would be required for a conventional stable platform tracking C_v.

3.3.2.5 Leveling alignment of an Inertial Navigation System platform.

During actual navigation, the computer can only "compute" where the C_v frame "should be." This may be labeled C_c, the computed NED local-level frame at the computed position (recall that the C_v frame axes are dependent on position). Both the platform system C_p and the computed axis system C_c are brought into close alignment with the C_v frame during the formal alignment process before flight (i.e., $C_p^v = C_p^c = I$). Thereafter, sensed angle misalignments between the computer C_c and the C_p frame acts as an error signal to the feedback system that drives the platform axes in the C_p frame.

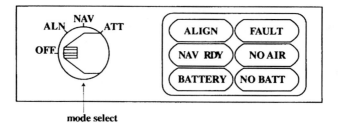

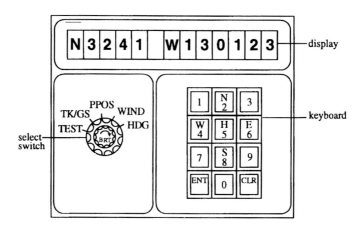

Fig. 3.14 Typical INS cockpit panels.

It is therefore important that the stable element be leveled accurately with respect to the local vertical and aligned in azimuth with respect to true North. Precise leveling of the stable element is accomplished prior to flight by the accelerometers that measure specific force in the horizontal plane. The stable element is reoriented until the output of the accelerometers is zero, indicating that they are not measuring any component of gravity and that the platform is level.

Leveling and azimuth alignment to true North is accomplished before flight by selecting ALIGN with the INS select switch as shown Fig. 3.14. This initiates the build-in test equipment (BITE) sequence after which the coarse alignment occurs. The annunciator ALIGN light illuminates at this time and may flash in conjunction with a FAULT light if aircraft motion is detected or if incorrect position information has been entered by the pilot. Note that if incorrect longitude is entered, the INS has no internal way to compensate but must rely on previously stored values for comparison. The entered latitude, however, must match the INS computed latitude at the end of the alignment.

Initial alignment consists of orientating the acceleration sensitive axes of the system within the navigation reference frame before it is required for use. There are various methods of accomplishing alignment, and the choice of methods depends on operational factors. Alignment methods can be classified as self-alignment, reference alignment, and alignments at sea and in the air. Transfer alignments are very important capabilities and will be discussed in the subsequent chapter on aided-INS.

Self-alignment normally consists of the following phases: warm up and coarse alignment, fine alignment, and gyrocompassing (discussed in the next section). The warm up and coarse alignment consist of rapid heating to operating temperature and gyro run up, leveling with reference to the aircraft frame and aligning to a preselected heading reference (such as a directional gyro or "best available true heading"—BATH heading). All phases can be carried out concurrently, but the time for the phase is ultimately limited by the platform heating (2–3 min) at moderate latitudes. Leveling and azimuth alignment by these means can be achieved to within a few arc minutes of 1 deg. At high latitudes a complete alignment may require 10 min or more for convergence.

3.3.2.6 Gyrocompassing alignment. A coarse gyrocompass alignment starts with the magnetic compass output and the application of variation to approximate the true North reference (expect accuracy within one 1 deg). From this point gyrocompassing is performed. This process makes use of the ability of the gyros to sense the rotation of the Earth. If the stable element is misaligned in azimuth, the East gyro will measure an incorrect Earth rate and will cause a precession about the East axis. This precession will cause the North accelerometer to tilt. The output of this accelerometer is then used to torque the azimuth and East gyro to ensure a true North alignment and a level condition. Because this alignment is dependent on Earth rate, it may be difficult to complete an accurate gyrocompass alignment rapidly unless an excellent true heading reference is passed to the unit.

Referring to Fig. 2.4, note that the Earth rate projection on the i_v axis of the local-level frame is $\omega_{e/i} \cos \lambda_e$ and zero on the j_v axis. If the i_p axis is misaligned in heading with the i_v axis by $\Delta\psi$, then the East axis gyro along the j_p axis will sense $\omega_{e/i} \cos \lambda_e \sin \Delta\psi$ and the platform will rotate about the j_p axis generating an output from the North accelerometer mounted on the i_p axis. This output is used to torque the azimuth gyro i_p axis back toward true North where $\Delta\psi = 0$, and equilibrium is achieved.

To summarize the alignment process, the local vertical is established by platform leveling. This is the most fundamental reference direction. To complete platform alignment, North must be known—this is accurately established by gyrocompassing. The INS is capable of gyrocompassing to an accuracy of 10 min of arc or less in 5 to 10 min. After the platform is aligned, it remembers its alignment and remains (for a reasonable length of time) pointing to true North and the local vertical regardless of the maneuvers of the aircraft.

3.3.3 Strapdown System (Analytic Inertial Navigation System Type)

It is possible, using digital computers and complex programming, to dispense with the gimbals and mount the inertial sensors directly on the airframe. Such a

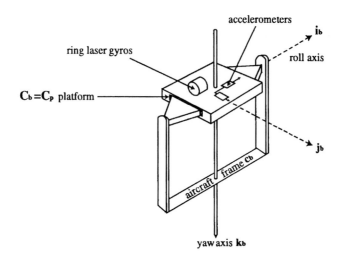

Fig. 3.15 Strapdown INS.

system is known as a strapdown INS and is shown in Fig. 3.15. The idea for an inertial navigation system using body-mounted sensors was described in a patent by W. Newell in 1956. However, such a system has only been practical since the development of small, powerful digital computers. Analog computers always could perform the required calculations in real time, but not accurate or reliable enough for inertial navigation.

In a strapdown system (SDS) the sensor package is bolted to the body axes, and the FEN is mechanized in C_b coordinates. Because the accelerometers are fixed to the vehicle, their orientations with respect to the local-level frame are defined by DCMs that are computed (this computation must be very rapid). Information for this rapid computation is obtained from the output of gyroscopes that are also directly mounted on the body of the vehicle that directly outputs $\omega_{b/i,b} = [P\ Q\ R]^T$. In effect, the computer in a SDS and the computed DCMs replace the gimbals in a conventional system. Advancements in digital computation, plus the rugged ring laser gyro, account for the widespread use of SDS systems.

Although the SDS is free from physical gimbal lock, the mathematical singularity still exists when computing the Euler angles ψ and ϕ at a pitch angle θ of ± 90 deg (see Sec. 3.1). A quaternion representation of rates within the computer avoids this; however, heading ψ and bank angle ϕ remain undefined and thus cannot be displayed in the cockpit at a pitch angle θ of ± 90 deg.

Two additional advantages are inherent in a SDS. Sensors can be in any relative position with respect to one another. This eases the assembly and servicing of SDS and enables the system to be updated as more recent sensors become available. Moreover, tracking the reference frame, including those orientation required for initial alignment, are not limited by physical inertia of the gimbals.

The accuracy of a SDS, like that of a platform INS, is usually limited by the accuracy of the inertial sensors. Error propagation is similar to that for platform systems, and discussion of this is deferred to Chapter 4.

3.3.4 Strapdown Mechanization

Assuming that the strapdown computation occurs in local-level coordinates, the fundamental equation is

$$\dot{v}_{k.v} = g_{\text{lv}.v} + C_b^v C_f^b a_{\text{sf}.f} - [\Omega_{(v/i.v)} + \Omega_{(e/i.v)}] v_{k.v} \quad (3.22)$$

The accelerometers are mounted in frame C_f that is permanently fixed in relation to the body axes C_b. The direction cosine matrix C_v^b can be determined from its derivative, which was derived in Sec. 2.5.1 and Eq. (2.39) as

$$\begin{aligned}
\dot{C}_v^b &= -\Omega_{b/v.b} C_v^b = -(\Omega_{b/i.b} - \Omega_{v/i.b}) C_v^b \\
&= -[\Omega_{b/i.b} - (\Omega_{v/e.b} + \Omega_{e/i.b})] C_v^b \\
&= -[\Omega_{b/i.b} - C_v^b(\Omega_{v/e.v} + \Omega_{e/i.v}) C_b^v] C_v^b \\
&= -\Omega_{b/i.b} + C_v^b[\Omega_{v/e.v} + \Omega_{e/i.v}]
\end{aligned} \quad (3.23)$$

Note the similarity transformation $\Omega_{v/i.b} = C_v^b[\Omega_{v/i.v}] C_b^v$ as in Eq. (2.35).

The rotation matrix (cross-product operator) for body axis rotation rates $\omega_{b/i.b} = [P\ Q\ R]^T$ is shown in Eq. (2.7), and the combined transport and Earth rate rotation matrix can be seen in Sec. 3.3.1 to be

$$-[\Omega_{v/e.v} + \Omega_{e/i.v}] = \begin{bmatrix} 0 & -\{\dot{\mu} + \omega_{e/i}\}\sin\lambda_e & \dot{\lambda}_e \\ \{\dot{\mu} + \omega_{e/i}\}\sin\lambda_e & 0 & \{\dot{\mu} + \omega_{e/i}\}\cos\lambda_e \\ -\dot{\lambda}_e & -\{\dot{\mu} + \omega_{e/i}\}\cos\lambda_e & 0 \end{bmatrix}$$

(3.24)

The Euler angles ($\psi\theta\phi$) can be found directly from the elements of direction cosine matrix C_v^b as given in Eq. (2.11) or from the body rates as given in Eq. (2.12). In practice quaternions (see Sec. 2.3.2) are used to avoid the mathematical singularity at a pitch angle of ± 90 deg (recall that heading ψ and bank angle ϕ are undefined at $\theta = \pm 90$ deg).

It is not necessary to use the computed local-level frame for the navigation solution as illustrated here. The primary advantages of a strapdown system are weight, size, cost, and reliability. These were offset in the past by the increased computation and loss of accuracy over systems using gimbals. Strapdown systems are subject to the same type of errors as local-level systems but have added errors due to in part to gyro torquing asymmetry, commutivity errors from using finite digital pulses for angles, and the normal truncation and round-off errors associated with computers. With the advance of digital computer technology and improved sensors like the ring laser gyro, however, the disadvantages have significantly decreased, and strapdown INS technology is now the INS technology of choice in aircraft systems.

3.3.5 Introduction to Inertial Navigation System Error Equations

Dominant INS errors are caused by imperfect knowledge of initial conditions (for example, those existing after alignment) and by error propagation in time.

The nine, nonlinear differential navigation equations—three from the FEN, three from integrating velocity to get position, and three from the equation for DCM rate of change—can be perturbed by a wide variety of error sources, not only those resulting from incorrect initial conditions. The perturbations of these equations, when kept small, can be shown to result in a linear set of differential equations. One set of these equations is called the Pinson error model, named after the man who derived it for space-stable mechanizations.

The interesting thing to remember is that, if the initial errors are known and if the errors are kept small, then the errors propagate deterministically and thus can be used to correct the INS output at any time during flight. Of course, eventually, gyro precession and system biases will dominate the outputs, and the errors will no longer remain small. Until then the INS outputs may be significantly reduced by applying the mathematical corrections from the error propagation equations. The equations may even be used to realign an INS system in-flight, but this is risky business because all useful output would be lost if the alignment failed.

Of course the initial condition errors are never known accurately (they would be taken out if they were). Remember that errors are both deterministic and statistical, and the statistical errors can only be estimated. Investigating the propagation of deterministic errors provides useful insight into system performance. There is such a wide variety of error sources, however, that it is extremely difficult to identify which ones are dominating the error response curves under all operating conditions.

This much, however, can be said: the velocity errors are excited by accelerometer errors (primarily bias and scale factor) and imprecision in knowing local gravity, and the attitude errors are significant due to gyro precession (which in itself is caused by many factors). Nearly all errors display common dynamic modes of oscillation, primarily the Schuler oscillation with the 84.4-min period and the 24-h error oscillation caused by Earth's rotation.

3.3.6 Mechanizing the Vertical Channel

If we investigate the vertical component of the equation

$$\int\int (a_{sf} - g_{lv}) \, dt \, dt \tag{3.25}$$

the time constant for the unstable mode will be 78 min, thus making this channel useless for precise altitude control. This instability is inherent *regardless of the mechanization* because local gravity g_{lv} is computed based on position. If the computer calculates an estimated position lower than the actual position, computed local gravity g_{lv} will be larger than it actually is and cause the computer to calculate a downward acceleration, resulting in a larger computed local gravity g_{lv} and so on. If the computer calculates an estimated position higher than the actual position, computed local gravity g_{lv} will be smaller than it actually is and cause the computer to calculate an upward acceleration, resulting in a smaller computed local gravity g_{lv} and so on. This aggravates the double integration that already exists and underlines the fact that the altitude channel must be bounded by an external independent source.

The barometric altimeter comes to the rescue here. It is consistently and reliably used to provide long-term stability (even though it is noisy and erratic in the short term). The mathematical combination of barometric and inertial altitude requires considerable analytical skill and much testing, but when successfully accomplished has proved accurate enough to use in demanding flight applications such as computing flight path and bomb trajectories.

One example of the way that inertial altitude may be constrained is given by Britting who defines

$$\hat{r}_k = (r_{\text{baro}})^b (r_{\text{inertial}})^{(1-b)} \quad (3.26)$$

where a nonlinear weighting of $b = 1$ selects the barometric altimeter exclusively and $b = 0$ selects the inertial. Filtered linear weightings provide another alternative. The specific technique will vary with the application and the mission and should be tested in-flight for verification.

The nonlinear choice of altitude for use in the navigation equations enters the gravity computation from its calculated value in the Earth-centered inertial frame C_i. Using harmonic mathematical expansion terms for the ellipticity of the Earth, Newtonian gravity in English units is given by Britting as

$$g_{m.i} = \frac{\eta}{(r_{\text{baro}})^b (r_{\text{inertial}})^{(3-b)}} \begin{bmatrix} k_{\text{eq}} & 0 & 0 \\ 0 & k_{\text{eq}} & 0 \\ 0 & 0 & k_{\text{pole}} \end{bmatrix} r_{k.i} = a(K_G) r_{k.i} \quad (3.27)$$

where

$$\eta \cong 1.407645(10)^{16} \frac{\text{ft}^3}{\text{s}^2}$$

$$r_{\text{baro}} = r_{\text{msl}} + h_{\text{baro}}$$

$$r_{\text{inertial}} = r_{\text{msl}} + h$$

$$r_{\text{msl}} = r_{\text{eq}}(1 - f \sin^2 \lambda_e)$$

$$r_k = \{(r_{\text{baro}})^b (r_{\text{inertial}})^{(3-b)}\}^{\frac{1}{3}}$$

$$k_{\text{eq}} \cong 1 + \frac{3}{2} \left(\frac{r_{\text{eq}}}{r_k}\right)^2 [1 - 5\sin^2(\lambda_c)] J$$

$$k_{\text{pole}} \cong 1 + \frac{3}{2} \left(\frac{r_{\text{eq}}}{r_k}\right)^2 [3 - 5\sin^2(\lambda_c)] J$$

$$\lambda_c = (\lambda_e - \delta_{\text{normal}}) \cong \lambda_e - f\left[1 - \frac{h}{r_{\text{eq}}(1 - f\sin^2\lambda_e)}\right] \sin(2\lambda_e)$$

$$J \cong 1.0823(10)^{-3} \quad \text{and} \quad r_{\text{eq}} \cong 20,925,696 \text{ ft}$$

The deviation of the normal σ_{normal} and many other parameters in the preceding equations were illustrated in Fig. 2.2 and defined in Eq. (3.12). The constant J

and equatorial radius r_{eq} are dependent on satellite observations and the harmonic expansion of the gravity field. Note that the weighting b between the barometric and inertial altitude as depicted here is a value between zero (use inertial altitude only) and three (use "baro" only).

If Eq. (3.26) were written in local-level frame C_v, it would become

$$g_{m.v} = C_i^v a K_G r_{k.i} = a(C_i^v K_G C_v^i) r_{k.v} \qquad (3.28)$$

where C_i^v was given in Eq. (3.13).

Regardless of the chosen coordinate system, barometric aiding introduces important considerations for flight applications that may be summarized as follows:

1) The inertial altitude is constrained in the long term.

2) The Central Air Data Computer (CADC) adjusts the baro altitude for atmospheric effects. During long climbs and descents, however, errors will build up in a nonstandard atmosphere (over 5% in some cases) because the CADC uses static pressure and because of biases in the altimeter itself.

3) Inertial navigation system indicated altitude has good short-term performance. However, biases in the system caused by maneuvering may take a few minutes to disappear after returning to level flight. This has significant influence on precision flight applications.

Remember: The Altitude Channel of an INS Requires Long-Term Aiding for Stability!

3.4 Attitude-Heading Reference Systems

An attitude-heading reference system (AHRS) is a set of inertial sensors that provides continuous attitude, heading, position, and velocity information. It is a compromise to a slaved directional gyrocompass system and a vertical gyroscope that provides low-accuracy position and velocity (but with unbounded errors). Advanced AHRS systems in use today employ strapdown RLGs and/or fiber-optic gyros. A magnetic flux device and accelerometers provide reference heading and specific force. It is strongly recommended that the AHRS be mounted near the center of gravity of an aircraft, especially if its outputs will be used for flight control.

In a typical AHRS, magnetic heading is assumed to be available for use as the heading reference, while the airspeed comes from the CADC and the pitot-static system. A Schuler-tuned vertical erection loop that uses true airspeed (TAS) to compensate for acceleration-induced errors is often used to reduce vertical error during maneuvering. Sensor information is merged to generate statistical estimates of aircraft attitude and heading. Navigation outputs are also available for use by other avionics systems. True airspeed controls the roll and pitch, as well as the AHRS alignment. The baro-altitude damps the system's vertical channel.

The integration of air data and inertial data allows wind, angle of attack, and angle of sideslip to be measured and displayed, possibly for use in an automatic flight control system (AFCS). Digital bus structures keep interfaces viable between the EFIS and the AFCS. Additional analog outputs provide synchro outputs for conventional aircraft instruments. It should be remembered that the AHRS is a dead-reckoning navigation device, susceptible to unbounded errors, and that it requires precise alignment for safe and proper operation.

3.5 Closure

During World War II the German V-1 missile was effectively employed against the British. It flew by holding attitude and altitude for a specific number of propeller rotations, then plunged destructively toward the Earth. It used a slaved directional gyro-compass and a vertical gyroscope to effectively accomplish its grim task. In the United States after the war, Charles "Doc" Draper at the Massachusetts Institute of Technology modified a 1933 invention of J. M Boykow, the gyroscopic pendulum, and made it into a practical integrating gyroscope. Soon he was flying across the country with a huge six-gimbaled inertial navigator in the cargo bay of a transport aircraft. By the late 1950s, however, his inertial navigator was being made small enough to fit into Navy attack aircraft that operated from a carrier. The local-level systems with analog electronics of the 1960s gave way to the digital local-level systems of the 1970s, along the way guiding astronauts to lunar landings and guiding submarines for months at a time in the depths of the ocean. The strapdown inertial navigation system of today, with its integrated digital electronics and advanced guidance capabilities, is one of the marvels of this century.

4
Concept of Uncertainty in Navigation

An inertial navigation system propagates two types of errors: deterministic and statistical. Mathematical models exist for statistical errors called *stochastic processes*. The deterministic and statistical errors, even though the Inertial Navigation System (INS) is a precision system, will eventually dominate, and the navigation output will become useless. These navigation outputs—position, velocity, and attitude (PVA)—therefore must be updated by measurements in order to be bounded. Schuler tuning automatically bounds many of the errors propagated in an INS and significantly enhances the long-term performance.

The first part of this chapter explains in detail how the INS deterministic errors propagate in time. The INS error model, a standard tool for many years, provides valuable insight into the inner workings of an INS. It will be apparent that an INS requires some type of aiding and updating in the long term to remain valid. The actual differential equations for INS operation are nonlinear, but the *error equations are valid for linearized versions of these differential equations*; hence, the requirement for the errors themselves to remain small, otherwise a linear analysis is not valid. A brief exposure to state-space dynamic modeling will be made because this technique is widely used in the aerospace community and is essential in understanding error propagation.

Statistical modeling will be introduced in this chapter by reviewing some basic concepts from probability and illustrating their applicability to navigation and flight testing. This will lead to a discussion of circular error probable (CEP) and, finally, to the batch processing of measurements that are at the heart of least-squares estimation. These tools will enable you to understand the importance of Kalman filtering as it applies to navigation.

This is important because the Kalman filter will have two strong impacts on system performance: it will "wash out" the adverse affects of not knowing the initial errors when the INS is initialized, and it will recursively combine different measurement updates that are aiding the INS to provide a most probable position (MPP) as accurate as that provided by a weighted least-squares batch solution.

After completing this topic, you should be able to do the following:

1) Explain the mathematical concept of position, velocity, and attitude error modeling and error propagation in time, why it is critically important in an INS, and how Schuler tuning bounds many of these errors so that the INS remains usable for extended periods in flight. Point out the significance of the Schuler cycle in testing inertial systems.

2) Generalize the concept of statistical modeling and uncertainty as indicated by expected value (average) and variance and extend the concept to multiple states (covariance matrix representation) that are applicable to aircraft flight and flight testing.

3) Relate expected value and variance to the test and evaluation dispersion measures of CEP, "distance root-mean-square" (drms as used by the FAA) and dilution of precision (DOP) errors.

4) Understand the concept of using external measurements to not only update INS outputs, but also to determine the full compliment of initial INS errors (even those that are not being directly measured) so that their effects can be eliminated (or reduced) during flight.

5) Recognize the fundamental equations of least-squares solution to multiple measurement updates and identify the practical value of a recursive solution to this problem.

4.1 Inertial Navigation System Error Propagation Equations

Standard derivations of error propagation equations are available for a wide variety of system mechanizations, although Schuler tuning makes many of the errors look similar in time regardless of the platform mechanics. These equations are typically too difficult to solve deterministically, and thus what is called an *error budget* is developed from multiple simulations of an INS system to be developed or tested. This error budget shows the statistical sensitivity of overall performance to system and sensor accuracy so that meaningful tradeoffs can be made during development and testing.

For example, the navigation outputs of the Standard INS are required to meet certain standards of accuracy. After a full gyrocompassing alignment, the standard medium accuracy INS must conform to the following published limitations on position error: 1) up to 1 h flight time: 0.8 n mile/h CEP; 2) up to 10 h flight time: 2.0 n mile /h CEP with 95% confidence; and 3) over 10 h of flight time: 20 n mile cross-track error and 25 n mile along-track error with 95% confidence. [Note that if the North and East position errors are assumed to be independent identically distributed zero-mean random processes, then the first two requirements are equivalent to a root-mean-square (rms) position error growth rate of 0.68 n mile/h per axis for the first hour of flight and 0.82 n mile/h per axis during the first 10 h of a mission.]

The preceding specifications should make sense to you by the time this chapter is finished, but it is stated here solely to illustrate the purpose of an error budget. The error budget would tabulate the sources of error during development or test and indicate the contribution of each error source to the total error. These contributions are not added like percentages up to 100%, but are root-sum-squared (RSS). That means each error contribution is squared, then all squares are added, and finally the square root is taken to predict the overall system error. By combining an error budget with a cost analysis, it would be apparent what sensor or system characteristic should be modified to meet the specification. An error budget makes good sense even if a specification has not been mandated.

4.1.1 Components of an Error Budget

Three types of error are important besides the distinction already made about deterministic (sometimes called systemic) and statistical (stochastic) errors. There are hardware component biases and scale factors, permanent misalignments of the sensors causing errors in reference frame orientation, and initial condition errors

CONCEPT OF UNCERTAINTY IN NAVIGATION 75

Table 4.1 Error budget table

Position error performance (breakdown by source)	Heading error performance (breakdown by source)
Initial condition error Gyro drift rates Accelerometer errors Gravity anomalies Math model complexity	Uncorrelated errors are root sum squared by column when determining performance for each requirement

(which may vary every time the unit is turned on). Discussion of the random component of these errors is deferred until later in this chapter.

Error models are determined for system parameters, inertial sensors, and external measurements. The models predict how each of these factors contributes to navigation output error (that is, position, velocity, and attitude). Such errors for attitude are not easy to define (see Sec. 4.1.2.1 on the δ_ψ equations that follow). Once defined, however, the relationship between the rms of these errors and the performance requirements may be charted in an error budget.

The matching of source errors to resulting performance errors in an error budget is valuable in evaluating system improvements, hardware and software changes, and algorithm selection (see Table 4.1). For example, if the performance error contribution from gyro misalignment is relatively small, specifications on that gyro may be reduced without a serious loss of performance. If the performance error contribution is large, it may be found that a specific gyro error source is dominant, thus focusing on the area that needs the most improvement. Using the error budget in this manner is often called a *sensitivity analysis*. One of the purposes of INS error modeling is to fill in the squares of the error budget table. How this is done is covered in subsequent discussion.

4.1.2 Inertial Navigation System Error Models

There are nine errors defined by a typical error model for an INS: three for position error, three for velocity error, and three for attitude error. This is called a "nine-state" model, and all states are grouped into a state-vector x. A matrix differential equation solves for x by finding a mathematical expression for each of the derivatives of x and mathematically integrating them from initial conditions under the action of forcing functions (the forcing funtions are grouped into a vector called u and weighted by control matrix B).

Specialized error models exist for particular mechanizations or simplifications. When the altitude and altitude-rate channels are neglected, a seven-state model results. The Pinson error model specifically defines attitude errors so that the effect of gyro drift, usually a dominant error source, can be isolated as a forcing function to the differential equations describing these errors. Regardless of the method of derivation, all INS error models rely on small perturbation theory to provide a coupled set of linear differential equations that may be used for simulation and

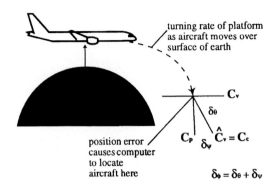

Fig. 4.1 The ψ equation for platform tilt.

analysis. Applying this theory to attitude errors is often done using the δ_ψ psi equations to provide added insight into system behavior. This process is discussed in the next section.

4.1.2.1 Psi equations.
To define error in attitude, suppose that an aircraft is actually located at the North Pole as shown in Fig. 4.1. If the INS had no error, then the local level platform axes C_p would be perfectly aligned with the true (ideal, error-free) axes $C_t = C_v$. But what if the INS computed the current position at 70° latitude? In the absence of any other errors except position, the computer would command the platform to track computer axes C_c, which would differ from C_t by 20°. A position error of this type causes the same type of platform misalignment that would exist from an error bias on the North accelerometer. Note that the errors discussed here are in practice a small fraction of a degree.

While the platform C_p was being torqued to align with C_c, two different error angles would exist. One is the 20-deg angle between computer axes and the ideal, error-free axes $C_t = C_v$, the unknown computer error δ_θ angle (the computer is unaware of this error); and one is the total angle error between the actual platform axes C_p and C_t called the total platform tilt error δ_ϕ angle. Note that there is another computed error δ_ψ angle as shown in Fig. 4.1 that the INS computer does know about and is trying to drive to zero (because the computer calculates that C_c is the local level plane and is unaware that the δ_θ error angle exists).

What is interesting here is that, even if the actual platform axes C_p and the idealized axes C_t were aligned with each other ($\delta_\phi = 0$) so that no tilt error existed at the actual North Pole position, the computer, calculating latitude to be 70°, would be aware only of the psi δ_ψ misalignment between platform C_p and C_c and would *adversely torque the platform away from the correct alignment.* Hence the term "psi equations," a term commonly applied to the error propagation equations for a local level INS. The computer will command a change in platform axes whenever the psi error exists.

Angles cannot be grouped into a vector unless they are small (less than 1 deg). This is because the order of rotation, vitally important for successive large angle rotations, is irrelevant when using vector components. Because it can be shown that order is irrelevant for small angle rotations (less than 1 deg), the following

vector relationship is valid:
$$\delta_\phi = \delta_\theta + \delta_\psi \tag{4.1}$$

The preceding is a vector equation. It is apparent from the preceding discussion that propagation errors are dependent on position and velocity errors as well as errors from sensors. Error analysis based on the psi equations results in the Pinson error model.

4.1.2.2 Mathematical formulation of small-angle attitude errors.
This section may be skipped without loss of continuity in the text. Applying small angle assumptions to Eq. (2.11) defining C_v^b, the definition of a DCM for a small, Euler-angle rotation vector $\delta_{b/v} = [\Delta\phi \ \Delta\theta \ \Delta\psi]^T$ is seen to be

$$C_v^b(\delta_{b/v}) = \begin{bmatrix} 1 & \Delta\psi & -\Delta\theta \\ -\Delta\psi & 1 & \Delta\phi \\ \Delta\theta & -\Delta\phi & 1 \end{bmatrix}$$

$$= [I - (\Delta t)\Omega_{b/v.b}]$$

$$= [I - (\Delta t)(\omega_{b/v.b}\times)] \tag{4.2}$$

It can be shown that for small angles the order of rotation is irrelevant in Eq. (4.2). This is necessary for $\delta_{b/v}$ to be a vector. Compare the result in Eq. (4.2) with Eq. (2.40) and note that $C_v^b(\delta_{b/v})$ may represent either a DCM for a small-angle vector rotation $\delta_{b/v}$ or a matrix factor used to compute a time-varying DCM from its matrix derivative.

To generalize the above result so that it is useful in error analysis, let the psi angles represent the misalignment angles between the platform C_p frame and the computed local-level frame C_{cv} so that

$$C_p^{cv}(\delta_\psi) = \begin{bmatrix} 1 & \delta_{\psi z} & -\delta_{\psi y} \\ -\delta_{\psi z} & 1 & \delta_{\psi x} \\ \delta_{\psi y} & -\delta_{\psi x} & 1 \end{bmatrix} = \{I - \Omega_\psi\}$$

$$= I - \Delta t \begin{bmatrix} 0 & -\dot\delta_{\psi z} & \dot\delta_{\psi y} \\ \dot\delta_{\psi z} & 0 & -\dot\delta_{\psi x} \\ -\dot\delta_{\psi y} & \dot\delta_{\psi x} & 0 \end{bmatrix} = \{I - \Delta t\Omega_{v/p.v}\} \tag{4.3}$$

Assuming small misalignments, the relationship between the computed local-level and the actual local-level reference frames is then a time-varying one given by

$$C_p^{cv} = \{I - \Omega_\psi\}C_p^v = \{C_p^{cv}(\delta_\psi)\}C_p^v \tag{4.4}$$

where $(-\Omega_{v/p.v})C_p^v$ by analogy may be considered the rate of change of computed C_p^{cv}. If small perturbation theory were used to generate

$$C_p^{cv} = C_p^v + (\delta C_p^v) \tag{4.5}$$

then the small perturbation (δC_v^p) is given by

$$(\delta C_p^v) = (-\Delta t\Omega_{v/p.v})C_p^v = -\Omega_\psi C_p^v \tag{4.6}$$

As an application of these equations, consider the output of the specific force sensors mounted on the platform. The computed specific force a_{csf} will have computed errors in the local-level frame according to the following relationship

$$a_{\text{csf}.v} = a_{\text{sf}.v} + \delta a_{\text{sf}.v} \tag{4.7}$$

This computed error will equal the small angle transformation of accelerometer errors on the platform times the platform errors, or

$$\begin{aligned} a_{\text{csf}.v} &= C_p^{cv}(a_{\text{sf}.p} + \delta a_{\text{sf}.p}) \\ &= \{I - \Omega_\psi\}C_p^v(a_{\text{sf}.p} + \delta a_{\text{sf}.p}) \\ &\cong C_p^v a_{\text{sf}.p} - \Omega_\psi C_p^v a_{\text{sf}.p} + C_p^v \delta a_{\text{sf}.p} \end{aligned} \tag{4.8}$$

By combining the preceding two equations,

$$\begin{aligned} \delta a_{\text{csf}.v} &= a_{\text{csf}.v} - a_{\text{sf}.v} \\ &\cong -\Omega_\psi a_{\text{sf}.v} + \delta a_{\text{sf}.p} \end{aligned} \tag{4.9}$$

The psi misalignment angles caused by gyro drift thus affect the NED components of accelerometer error through the preceding equations (neglecting higher order terms).

4.1.2.3 State-space formulation.

If we include the altitude channel, the Pinson error model will have nine states. These states are arranged into a state-space system of general form

$$\dot{x} = Ax + Bu \tag{4.10}$$

where the system is arranged so that the first derivatives of the state vector x are on the left and where u designates a vector forcing function. Matrices A and B depend on the system and error forcing function dynamics. For a typical INS error model, errors are relative to the mechanized navigation axes C_n. Usually the horizontal bases of C_n are set at a wander angle to the bases of C_v for a local level mechanization. The states in the state vector x are the three components of position error (errors in latitude, longitude, and height), the three components of velocity error, and the three components of the psi misalignments between the computer and platform axes.

A complete derivation of the error equations is beyond the scope of this text, and only a broad outline will be given here. For simplicity the altitude channel will be constrained. Recall that the computer is aware only of the psi δ_ψ misalignments between platform C_p and C_c coordinate frames. Platform drift (caused by gyro drifts) will cause a discrepancy between what the computer calculates the platform orientation to be (depending only on position) and what the orientation of C_p actually is. This drift will not cause any misalignment between computer-calculated orientation and the true axes because the misalignment δ_θ is presumed only caused by an error in position. This implies that

$$\dot{\delta}_\psi = \varepsilon \tag{4.11}$$

where the vector ε is the net drift due to gyro drift and torquing errors. Because we are dealing with drifts with respect to C_i, the Coriolis theorem must be applied to express the equation in computer axes, thus

$$\frac{d_i}{dt}\delta_\psi = \frac{d_c}{dt}\delta_\psi + (\omega_{c/i} \times \delta_\psi) = \varepsilon \quad (4.12)$$

which when expressed in C_c coordinates is

$$\dot{\delta}_{\psi.c} \cong -(\omega_{c/i.c} \times \delta_{\psi.c}) + \varepsilon_{\text{gyro}} \quad (4.13)$$

The drift ε has been expressed in the axis system where the instrumentation is located and transformed into the computer frame. The drift will become the forcing function $\boldsymbol{u}$ in the state-space Eq. (4.10). The rest will be absorbed into the $\boldsymbol{A}$ matrix.

The $\boldsymbol{A}$ matrix of the state-space system mathematically defines the inherent, unforced coupling between the error states. The $\boldsymbol{B}$ control matrix, on the other hand, mathematically depicts the relationship between the error source values and their mapping into navigation performance errors. Siouris presents the following simplified error equations for a local-level inertial navigation system mechanized in local-level coordinates, where ω_s is the Schuler frequency:

$$\delta\dot{\boldsymbol{r}}_{k.v} = \begin{bmatrix} 0 & -\{\dot{\mu}\}\sin\lambda_e & \dot{\lambda}_e \\ \{\dot{\mu}\}\sin\lambda_e & 0 & \{\dot{\mu}\}\cos\lambda_e \\ -\dot{\lambda}_e & -\{\dot{\mu}\}\cos\lambda_e & 0 \end{bmatrix} \delta\boldsymbol{r}_{k.v} + \delta\boldsymbol{v}_{k.v}$$

$$\delta\dot{\boldsymbol{v}}_{k.v} = \begin{bmatrix} 0 & -\{\dot{\mu}+2\omega_{e/i}\}\sin\lambda_e & \dot{\lambda}_e \\ \{\dot{\mu}+2\omega_{e/i}\}\sin\lambda_e & 0 & \{\dot{\mu}+2\omega_{e/i}\}\cos\lambda_e \\ -\dot{\lambda}_e & -\{\dot{\mu}+2\omega_{e/i}\}\cos\lambda_e & 0 \end{bmatrix} \delta\boldsymbol{v}_{k.v}$$

$$+ \begin{bmatrix} \omega_s^2 & 0 & 0 \\ 0 & \omega_s^2 & 0 \\ 0 & 0 & 2\omega_s^2 \end{bmatrix} \delta\boldsymbol{r}_{k.v} + \begin{bmatrix} 0 & \delta\psi_z & -\delta\psi_y \\ -\delta\psi_z & 0 & \delta\psi_x \\ \delta\psi_y & -\delta\psi_x & 0 \end{bmatrix} \boldsymbol{a}_{\text{sf}.v} + \delta\boldsymbol{a}_{\text{sf}.p}$$

$$\dot{\delta}_{\psi.c} = \begin{bmatrix} 0 & -\{\dot{\mu}+\omega_{e/i}\}\sin\lambda_e & \dot{\lambda}_e \\ \{\dot{\mu}+\omega_{e/i}\}\sin\lambda_e & 0 & \{\dot{\mu}+\omega_{e/i}\}\cos\lambda_e \\ -\dot{\lambda}_e & -\{\dot{\mu}+\omega_{e/i}\}\cos\lambda_e & 0 \end{bmatrix} \begin{bmatrix} \delta\psi_x \\ \delta\psi_y \\ \delta\psi_z \end{bmatrix}$$

$$+ \varepsilon_{\text{gyro}.p} \quad (4.14)$$

The error forcing functions are considerably simplified in the preceding equation. The gyro and accelerometer errors often are modeled with differential equations of their own, and this adds to the order of the state matrix $\boldsymbol{A}$ that becomes augmented with the new dynamics.

Britting and other authors have performed error analysis using the total error δ_ϕ between the platform and the local-level frame. Using the LaPlace operator, a compact form results for the local-level error equations, shown here with gyro drift and accelerometer bias forcing functions:

$$\begin{bmatrix} s & \omega_{e/i}\sin\lambda_e & 0 & \omega_{e/i}\sin\lambda_e & (-\cos\lambda_e)s \\ -\omega_{e/i}\sin\lambda_e & s & -\omega_{e/i}\cos\lambda_e & s & 0 \\ 0 & \omega_{e/i}\cos\lambda_e & s & \omega_{e/i}\cos\lambda_e & (\sin\lambda_e)s \\ 0 & -g & 0 & r_k s^2 & (r_k\omega_{e/i}\sin 2\lambda_e)s \\ g & 0 & 0 & -(2r_k\omega_{e/i}\sin\lambda_e)s & (r_k\cos\lambda_e)s^2 \end{bmatrix}$$

$$\cdot \begin{bmatrix} \delta_{\phi X}(s) \\ \delta_{\phi Y}(s) \\ \delta_{\phi Z}(s) \\ \delta\lambda_e(s) \\ \delta\mu(s) \end{bmatrix} = \begin{bmatrix} \text{gyro drift}_{\text{NORTH AXIS}}/s \\ \text{gyro drift}_{\text{EAST AXIS}}/s \\ \text{gyro drift}_{\text{DOWN AXIS}}/s \\ a_{\text{sfNORTH}}/s \\ a_{\text{sfEAST}}/s \end{bmatrix} \qquad (4.15)$$

which produce solutions of the form $(1 - \cos\omega_s t)$ modulated at the Schuler frequency.

It is apparent that the solution of the error equations typically requires a digital computer. For each error source and given set of initial conditions, each output navigation error will have a trajectory defined by the state-space model. When all of the error sources act together, as they certainly do in actual flight operations, the net output navigation error is a sum of all of the trajectories from each error source. This can be very complex to analyze, but it is the purpose of inertial guidance testing to generate these error plots. Analysts then practice reverse engineering to identify the dominant error sources and estimate their value from the test data. To do this, they use a mathematical tool called "regression analysis."

A typical set of solutions for latitude error resulting from a variety of error sources is shown in Fig. 4.2. It is easy to spot the Schuler mode of oscillation with an 84-min period. If all error sources were acting at the same time, the sum of all of these plots would provide a net position error plot for one flight test. A set of such plots, overlaid on the same axes, would summarize a set of flight tests, and it would be left to the program analyst to draw a thick line through the entire set from which the nautical miles per hour performance would be determined and checked against the system requirements. The way that this is done in practice is covered in the data reduction section of Chapter 9.

Perhaps it is now understandable why certain INS systems, based in certain locations, always shut down with similar errors (such as reading 5 ft per second when stopped in the chocks). If the error sources for the systems are similar, and if the flights are approximately the same duration from the same location, then the error at shutdown will be similar. It is also understandable how aircraft with redundant inertial navigation systems can have widely varying position and velocity readouts during certain portions of the flight and good agreement at other times. The error depends on where the INS computes its position and velocity relative to its Schuler oscillation.

If error plots are consistently repeatable for a given INS, then the error can usually be taken out by systems calibration in the lab. However, while in flight, error sources such as gyro drift and accelerometer scale factor propagate in a

CONCEPT OF UNCERTAINTY IN NAVIGATION

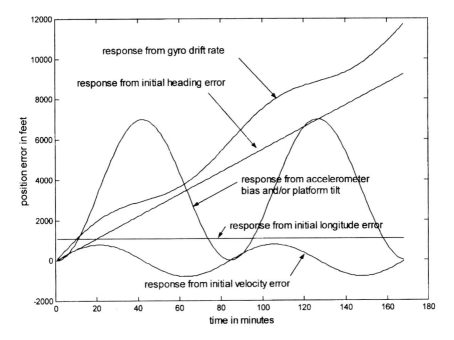

Fig. 4.2 Position error plots from multiple error sources.

random fashion. Thus, to be neutralized while flying, they must be estimated while flying. This is one of the jobs of the Kalman filter discussed in the next chapter.

Remember that the goal of these error plots is to make an error budget (see Table 4.2) for use in sensitivity analyses. For complex aided navigation systems, 40 or more error sources may be modeled in the state-space formulation.

It should be emphasized here that what keeps us from taking out the error sources is not knowing their values throughout the generation of an error trajectory, combined with not knowing the initial state errors at the start of the mathematical

Table 4.2 Gyrocompassing performance (steady state example)

Error source	Heading misalignment, arc min, rms
x-axis tilt error (0.01 deg/h rms)	3.1
y-axis tilt error (0.01 deg/h rms)	0.24
East velocity error (3 ft/s rms)	9.3

82   INTEGRATED NAVIGATION AND GUIDANCE SYSTEMS

integration. If these were known, most if not all of the deterministic error could be removed from the navigation outputs while the aircraft is flying.

Deterministic error, however, is only one type of error. The other type is statistical in nature. The way in which statistical errors are treated in modern navigation systems is the next topic.

4.2 Probability Concepts in Navigation

The need for statistical concepts to describe errors are easily illustrated with the plot in Fig. 4.3 that shows the simulated deflection of the vertical (the difference between local gravity direction and the normal to the Earth's ellipsoid) at various longitude locations. In our discussion of error models for INS systems, two such deflections were defined, one in the meridian plane and one out of the meridian plane (see Fig. 2.2). These deflections would be an error source during flight. Now, if a state-space representation of the errors were used with this deflection as an error forcing function, the deflection would have to be known and stored before flight for all possible flight locations. Because this is usually impractical, statistical measures are used to represent the data.

This type of plot exists for each and every error source that excites the error model of the state-space system. Thus the navigation output errors will behave differently for every flight as well. What is needed here is a concept of uncertainty that is practically meaningful for describing such system behavior and its associated data in test, development, and analysis. This "practically meaningful uncertainty

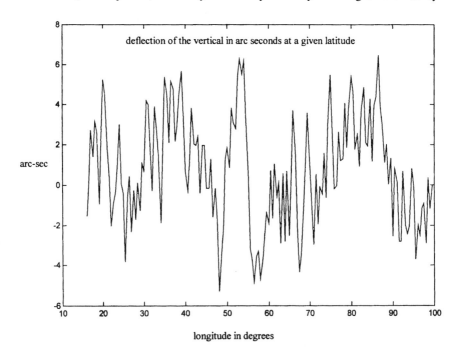

Fig. 4.3   Deflection of the vertical versus longitude.

concept" as presented here can be divided into statistical measures for test (CEP and radial rms distance) and probabilistic concepts and methods for both test and engineering analysis (Gaussian distributions, Bayes' theorem, random variables, stochastic processes, covariance, autocorrelation, spectral density, and most probable position).

4.2.1 Circular Error Probable

The dispersion of a navigation error can be estimated by investigating a large number of two-dimensional data sample pairs (for example, as position error along-track and position error cross-track). Given the mathematical properties of ergodicity and stationarity (defined later in Sec. 4.2.3.2), the circular error probable (CEP) is the statistical estimate of the radius of a circle that encloses 50% of the sample population (population implies the set of test data results). To obtain the CEP from two-dimensional data distributions, this section presents the relationships derived by Kayton and Fried.

To be mathematically rigorous, the data distributions must meet important mathematical constraints (bivariate Gaussian). The practical implications of this require that, given a two-dimensional data set, the principle data axes C_d orientation be found from the origin of the *mean error point* on the ground. For example, suppose the two-dimensional data set is expressed in pairs relative to the origin of C_v coordinates (North-East-Down). The heading angle Ψ required to find the principle axes C_d orientation is given by

$$\tan 2\Psi = \frac{2\sigma_{\text{north}}\sigma_{\text{east}}}{\sigma_{\text{north}}^2 - \sigma_{\text{east}}^2} \tag{4.16}$$

which is calculated using the estimates of standard deviations "squared" along C_v axes

$$\sigma_{\text{north}}^2 = \frac{\Sigma(\text{data}_{\text{north}})_i^2}{N} \tag{4.17}$$

and

$$\sigma_{\text{east}}^2 = \frac{\Sigma(\text{data}_{\text{east}})_i^2}{N} \tag{4.18}$$

and the correlation coefficient ρ from

$$\rho = \frac{\Sigma(\text{data}_{\text{north}})_i(\text{data}_{\text{east}})_i}{\sigma_{\text{north}}\sigma_{\text{east}}} \tag{4.19}$$

With the new principle axes C_d orientation, the data pairs are reconstructed and the standard deviations recalculated along the principal axes (using the formulas as in the preceding equation). If these standard deviations are denoted σ_1 and σ_2, then along the appropriate principle axis 68.3% of the data points are within one sigma (95.4% within two sigma) of the mean in the appropriate direction for σ_1 or σ_2. If σ_1 and σ_2 meet the criteria that $\sigma_1/3 < \sigma_2 < 3 < \sigma_1$, then the CEP is

$$\text{CEP} = 0.59(\sigma_1 + \sigma_2) \tag{4.20}$$

Note that these statistics, including the mean error, do not carry much technical information that is useful for analysis. They are valuable, however, for testing competing systems or for testing against specifications. A final note (possibly useful for test) is that if σ_1 and σ_2 are approximately equal, the CEP is approximately 1.18 times σ_1 or σ_2, and 95% of the data set points are within a circle of radius 2.45 times σ_1 or σ_2.

Summarizing the most important point, CEP, a radius enclosing 50% of the statistical population, is a statistic that measures dispersion; *it is not the mean* or average position error, but rather the dispersion of data samples about the mean. A small CEP means that all of the aircraft and missiles went to the same place, but that place was not necessarily the desired destination or target.

4.2.2 Twice the RMS Radial Distance Error

If two-dimensional pairs of source errors in a navigation system have a Gaussian distribution, then the 2 drms (twice the rms radial distance error) value is the radius of a circle about the true position containing approximately 95% to 98% of the statistical population of errors. As the bivariate elliptical distribution for the data becomes more circular (elliptical), the 2 drms scatter tends to 98% (95%) of the statistical population. This measure of accuracy is a statistic used extensively by the Federal Aviation Administration for all phases of flight testing and systems development, including air traffic control. In one dimension 2 drms is the 95th percentile, or 2σ.

4.2.3 Probability Concepts Important for Error Analysis

4.2.3.1 Probability applied to one variable.
A probabilistic event (ξ) is one to which a value can be assigned. A random variable is a function or assignment of value that maps probabilistic events into real numbers. For example, suppose there are four INS test flights (events ξ_1 through ξ_4) that result in final position errors of 1 n mile, 2 n miles, 3 n miles, and 4 n miles, respectively. The four position errors are realizations of the random variable X that maps the events (ξ = flights) into real numbered values x (such as final position error).

Probability theory answers the following questions about this set of test data: 1) What is the probability that a flight (ξ event) occurred where the position error (X random variable) was less than 1 n mile ($x < 1$)? 2) What is the probability that a flight (event) occurred where the position error (random variable) was less than 2 n miles ($x < 2$)? 3) Less than 3 n miles ($x < 3$)? 4) Less than 4 n miles ($x < 4$)? 5) Equal to or greater than 4 n miles? The answers to the preceding questions for the four sample test flights discussed in the preceding are 1) 0; 2) 1/4; 3) 2/4; 4) 3/4; 5) 1.

The answers to the above questions constitute $F(X)$, a function of random variable X that when plotted versus real number x is called the *probability distribution function*. This familiar plot of $F(X)$ versus real number x is shown in Fig. 4.4 where the preceding above questions are expressed in mathematical terms using $p(\xi_{X<x})$, the probability of an event ξ occurring where random variable X is less than real number value x. Real number x may be positive or negative ($-\infty < x < \infty$).

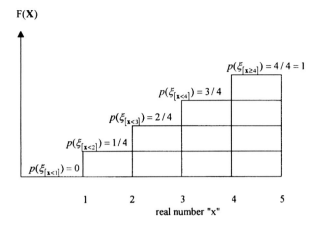

Fig. 4.4   Probability distribution function given four events.

The expected value (average) of position error is given by the weighted sum

$$E(X) = \Sigma_\xi x p(\xi_{X<x}) = \frac{\Sigma x_i}{N} \quad (4.21)$$

which for the four flight examples in the preceding paragraphs gives the average or mean as

$$E(X) = 1(1/4) + 2(2/4) + 3(3/4) + 4(4/4) \text{ n mile}$$

$$= \frac{1+2+3+4}{4} \text{n mile} = 2.5 \text{ n miles} \quad (4.22)$$

and an unbiased estimate of the variance (square of standard deviation) may be found using

$$\sigma^2 \approx \frac{\Sigma[x_i - E(x_i)]^2}{N-1} \approx 1.67 \text{ n miles}^2 \quad (4.23)$$

Gaussian "normal" distribution.   If the number of flights in our preceding example were very large and if they were normally distributed, then the cumulative distribution function would be called *Gaussian*. The derivative of this function, the probability density function (pdf), would be a *normal Gaussian* pdf given by

$$p(x) = [1/\sigma \sqrt{2\pi}] \exp -\left\{-\frac{[x - E(x)]^2}{2\sigma^2}\right\} \quad (4.24)$$

where $E(x)$ is the mean m and σ^2 the variance, or square of the standard deviation (SD):

$$E(X) = \int_{-\infty}^{\infty} x p(x) \, dx \quad (4.25)$$

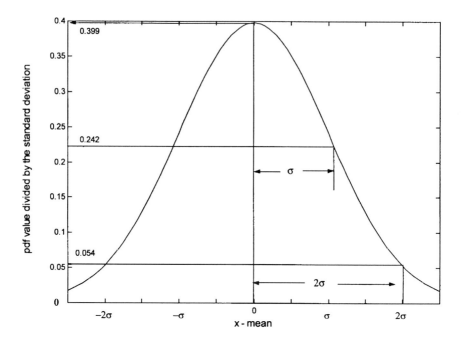

Fig. 4.5 Gaussian probability density function.

$$\sigma^2 = \int_{-\infty}^{\infty} [x - E(x)]^2 p(x)\,dx = E(X^2) - [E(X)]^2 \quad (4.26)$$

The familiar plot of the normal pdf is shown in Fig. 4.5 where the area under the curve between plus and minus one sigma represents 68.3% of all possible events ($2\sigma = 95.45\%$, $3\sigma = 99.73\%$). Although normal distributions commonly occur only for large numbers of uncorrelated events (Central Limit theorem of probability), the properties of these distributions for small sample data sets are extensively used in practice.

Bayes theorem. Given two events A and B and pdf p for each

$$p(A + B) = p(A) + p(B) - p(A * B) \quad (4.27)$$

with conditional probability $p(A/B)$ such that

$$p(A * B) = p(B)p(A/B) \quad (4.28)$$

Bayes theorem states that

$$p(A/B) = p(A)p(B/A)/p(B) \quad (4.29)$$

Assuming that the data is Gaussian, this result may be used to derive the Kalman filter, which provides a most probable estimation of navigation states given a set of measurements whose probability distributions are known. In the next chapter the Kalman filter is derived with an alternative method.

CONCEPT OF UNCERTAINTY IN NAVIGATION 87

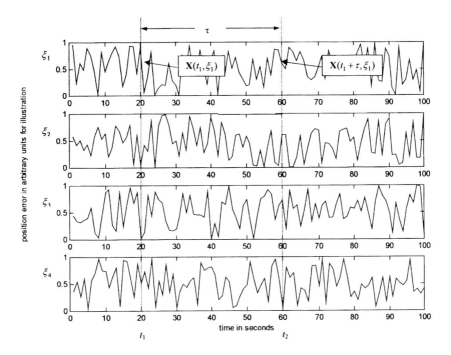

Fig. 4.6 Stochastic process given four events.

4.2.3.2 Stochastic process. This concept can best be explained by returning to the example of the four INS test flights. Suppose there are four INS test flights (events ξ_1 through ξ_4) that result in four position error trajectories in time. Note that the events trigger time functions in this scenario whose values at a specified time t_1 are the position error data used in the original example. Each of the time function plots in Fig. 4.6 is a *realization of a stochastic process* (position error in time) for random variable X. The realizations of random variable X at time t_1 are the values of each process at that time, shown in the figure by the intersection of the process and the first vertical dashed line. At time t_1 the probability distribution for random variable X, if it is Gaussian, will look similar to Fig. 4.5.

Note that one time plot alone will not allow the computation of expected value or variance because the random variable as defined is averaged across events and not along time. An ergotic process by definition implies the same mean will result from a time average "in the limit." Probabalistic measures of mean and variance, as defined here, are not test statistics (like CEP) that are obtained from individual test data points. The variance of a stochastic process should be estimated from a set of stochastic processes and thus may vary in time. When a large set of simulated stochastic processes are generated in a computer, a process called a *Monte Carlo simulation*, the mean and variance of the pdfs of the stochastic processes may be computed as time-varying functions. If these values are constant in time, the stochastic process is *stationary*.

Autocorrelation and spectral density. There are two more properties of a stochastic process that are very important in practical applications, the *autocorrelation function* and its Fourier transform, the *power spectrum density function* (variance spectrum). For our purposes, and referring to Fig. 4.6, the autocorrelation function $R(\tau, t_1)$ is given by

$$R_x(\tau, t_1) = E[X(t_1)X(t_1 + \tau)]$$
$$\approx \frac{X(t_1, \xi_1)X(t_1 + \tau, \xi_1) + \cdots + X(t_1, \xi_4)X(t_1 + \tau, \xi_4)}{N = 4} \quad (4.30)$$

where once again the averaging across event space is apparent for a particular value of time difference τ. If the stochastic process is wide sense stationary, then the same value for $R_x(\tau)$ will result no matter what start time t_1 is used. This is important for two reasons. First, it means that the stochastic process functions do not change in time (the start and end points are thus arbitrary) but are only dependent on the time difference τ; and second, that the Fourier transform of $R_x(\tau)$ is the power spectral density (psd) Φ_x given by

$$\Phi_x = \int_{-\infty}^{\infty} R_x(\tau) e^{-j\omega\tau} \, d\tau \quad (4.31)$$

which shows how the variance of the process is distributed with frequency.

Practically speaking, this means that if the autocorrelation function is slowly changing with τ, implying sinusoidal characteristics, the power spectral density will have a "spike" at the oscillation frequency. On the other hand, if the plots from a stochastic process are very erratic (like the hiss from a radio receiver), the process is referred to as "white noise" and the autocorrelation function will be near zero for all time differences τ except $\tau = 0$ (see Fig. 4.7). The psd for this process versus frequency will be flat. If real data shows this characteristic, it has high energy (if the psd stayed flat for all frequencies, no matter how high, the process would have infinite energy; thus, all real processes will have a limited *bandwidth*).

Probability applied to multivariable analysis. If plots of velocity error are also available, then another stochastic process for random variable Y may be analyzed alongside those for position error X using time as a common parameter. The autocorrelation function is defined in a similar way except one of the random variables X at a specified time is replaced with the random variable Y. The cross-correlation function results, and the correlation coefficient is defined as

$$\rho_{xy} = \frac{E[xy] - E(x)E(y)}{\sigma_x \sigma_y} \quad (4.32)$$

which is again computed across all events at a particular time.

This idea can be extended to multivariable systems, such as the PVA outputs of a navigation system. Autocorrelation and cross-correlation functions may now be placed into a matrix, the correlation matrix R_x, with autocorrelation functions for each random variable on the diagonals and symmetric cross-correlation functions off-diagonal. Note that a mixing of units occurs for cross-correlation functions off the diagonal. With this background, one of the most important practical concepts from stochastic processes may now be presented.

CONCEPT OF UNCERTAINTY IN NAVIGATION

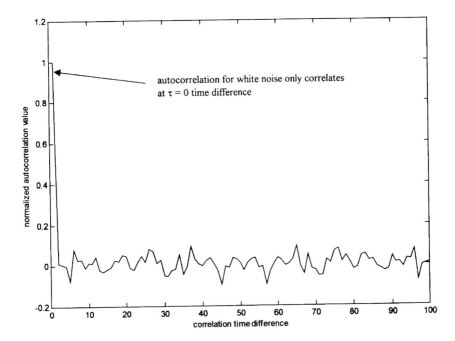

Fig. 4.7 Auto correlation function for white noise.

4.2.3.3 Covariance matrix P. Much of the probability theory to this point has been presented to prepare the way for this most important concept. The covariance matrix P and its propagation in time are vital in both describing and analyzing physical test results and comparing them to theoretical predictions. Error budgets for nearly all costly, complex systems involve understanding the covariance matrix P.

The covariance P (not a matrix yet) between two random variables is defined similarly to the cross-correlation. For two random variables X_1 and X_2 from the same process,

$$P_{X_1 X_2} = E\{[X_1 - E(X_1)][X_2 - E(X_2)]\} = E[(X_1)(X_2)] - E(X_1)E(X_2) \quad (4.33)$$

which is valid for a particular time (not indicated in the preceding equation). If the two random variables are made into components of vector x, then the covariance matrix P is given by

$$P = E[xx^T] = \begin{bmatrix} P_{X_1 X_1} & P_{X_1 X_2} \\ P_{X_2 X_1} & P_{X_2 X_2} \end{bmatrix} \quad (4.34)$$

It should be straightforward to extend the idea of a covariance matrix for more than two random variables. This concept is ideally suited for the state-space vector x whose components consist of important system performance variables.

In the case of the INS error states for position, velocity, and attitude, P is a nine-by-nine matrix. The nine variances of position, velocity, and attitude errors are on the diagonal and the remaining covariances relating the differing states are off the diagonal and symmetric. How this matrix propagates in time is of great interest, for the variances, if large or unbounded, will soon degrade INS performance. Position or velocity fixes are made to reset the values in the covariance matrix as will be discussed later. For a stochastic process made up of navigation error states that are largely zero mean, the variances reduce to the mean square moment of the random variable.

Note that the sum of the diagonals of this matrix is the root-sum-squared of all of the variances in the stochastic process. This number is a significant indication of the overall performance of the system and often is used as a "metric to minimize" when making decisions about how to treat navigation data. *The sum of all the elements on the diagonal is called the trace of the covariance P matrix.* Finally, note the physical significance of P and realize that the diagonals must always be positive if they denote variances. Mathematicians add a further restriction and state that P must be *positive definite* (PD). A cross-variance matrix between similarly dimensioned vectors x and w [$P_{xw} = E(xw^T)$] also may exist.

4.2.3.4 Covariance propagation in time.

It should be apparent why the propagation of covariance P is so vitally important for an INS that is fundamentally dead-reckoning its way between updates. If a computer is implemented to solve the state-space error equations, then it also can propagate the covariance matrix of those errors. This is accomplished nearly continuously with a computer so that error estimates and covariances are available at the discrete time t_k when measurements occur.

To propagate the covariance matrix during actual flight, the navigation errors must first be put into the state-space system given in Eq. (4.10). Then the initial uncertainty of the states must be known (or approximated), and the uncertainty of each of the error sources must be known (or approximated) as a variance throughout the flight. For example, if gyro drift ε has three components for drift (about each axis), then the matrix covariance of gyro drift ε (usually called the Q matrix) must be known or approximated. This is a three-by-three matrix that at time t_k is given by Q_k where

$$Q_k = E(\varepsilon_k \varepsilon_k^T) \qquad (4.35)$$

Error source uncertainties like gyro drift ε are often modeled as a white noise vector process v_k weighted by a matrix Γ_k, so $\varepsilon_k = \Gamma_k v_k$. If the covariance of v_k is Q_{sk} then $Q_k = (\Gamma_k Q_{sk} \Gamma_k^T)$. The important point here is that the covariance of the driving noise Q_k that enters the state equation must be known or approximated.

The state equation must also be put into its discrete form at time instant t_k. The discrete form of Eq. (4.10) presented without derivation is

$$x_{k+1} = A_k x_k + B_k u_k + w_k \qquad (4.36)$$

where all of the error source uncertainties have been lumped into w_k (gyro drifts, accelerometer noise, etc.), and where $x_{k+1} = x(t_{k+1})$. This is now a *difference* equation as opposed to a *differential* equation [it can be found from Eq. (4.10) in a straightforward manner by letting $\dot{x} \cong (x_{k+1} - x_k)/\Delta t$]. The deterministic forcing term $B_k u_k$ does not effect the covariance propagation equation.

Finally, before presenting the covariance propagation equation, a symbol is required to denote state error values and covariance values that exist before and after update, assuming that the update will occur at time t_k. This is done by denoting the covariance matrix before update at time t_k by $P_k(-)$ and after update at time t_k by $P_k(+)$. Remember, the computer "time tags" the data even though time in real life flows on continuously, and this notation indicates that the covariance matrix values before and after update will have the same time tag as assigned by the computer.

The covariance of the navigation error states P propagates according to

$$E(x_{k+1}x_{k+1}^T) = E(A_k x_k + B_k u_k + w_k)(A_k x_k + B_k u_k + w_k)^T$$

$$E(x_k w_k^T) = 0 \quad \text{(assumption)}$$

$$P_{k+1}(-) = A_k P_k(+) A_k^T + Q_k$$

$$Q_k = E(w_k w_k^T) \tag{4.37}$$

Equation (4.37) indicates that the error state covariance after updating at time t_k is propagated by the state matrix and the noise covariance Q_k of the error sources before updating at t_{k+1}. Updating in this case refers to applied measurements.

Note that if the value of the covariance changes with time, then the stochastic process describing the errors cannot be wide sense stationary in a probabilistic way. This occurs because inertial systems have undamped oscillatory characteristics (at the Schuler frequency) and because the linearized error equations are time varying. Although the navigation errors are not stationary, the rms navigation output performance measures for a stated length of flight time are often directly associated with a rms input value from a given error source as in Table 4.1. The resulting net performance error from all error sources is thus considered a root-sum-squared value, often given as nautical miles of error per hour. This presumes a linear propagation of covariance for the errors instead of the propagation obtained using Eq. (4.37).

4.3 Least Squares

When confronted with multiple sources of information at the time of update, a pilot in flight must determine which measurements are the most reliable and how they will be processed. Navigation systems must accomplish the same task automatically. The least-squares solution to this problem is well known and will not be formally derived here. By way of an example, however, the solution to the least-squares problem will be intuitively developed and discussed.

Suppose that a navigation system at time t_k has three potential updates to its latitude error that are stored in a vector z_k. Suppose further that the computer stores an internal estimate of latitude and tilt error in a vector $\hat{x}_k$ where the "hat" indicates an estimate (as opposed to the actual value). Finally, let the estimated measurement error be given by $\hat{z}_k = H_k \hat{x}_k(-)$ where $\hat{x}_k(-)$ is the error estimate before processing the update measurement and H_k is a linearized mathematical relationship valid at time t_k.

After the update using the actual measurement error z_k, the error estimate will be designated $\hat{x}_k(+)$. The method in which the measurement is linearized will be covered in detail in Chapter 6, but for now assume the following:

1) For the GPS system, suppose that the measurement error at time t_k has been linearized in some fashion and is given by the following relationship

$$z_{1k} = 1.1\hat{x}_{1k}(-) + 1.1\hat{x}_{2k}(-) + \varepsilon_1 \qquad \text{GPS (4.38)}$$

2) For the TACAN system, the linearized measurement error at time t_k is

$$z_{2k} = -2.2\hat{x}_{1k}(-) - 2.8\hat{x}_{2k}(-) + \varepsilon_2 \qquad \text{TACAN (4.39)}$$

3) For the PILOT, who estimates his or her error by looking out the window, suppose that the error in the estimate is given by

$$z_{3k} = 0.0\hat{x}_{1k}(-) + 0.5\hat{x}_{2k}(-) + \varepsilon_3 \qquad \text{PILOT (4.40)}$$

Suppose further that the pilot estimates his or her accuracy to be plus or minus one degree mean-squared error. If it is known that the TACAN is 10 times more accurate than the pilot and the GPS 10 times more accurate that the TACAN, the preceding three equations may be put in a weighted matrix form

$$\begin{Bmatrix} 100 & 0 & 0 \\ 0 & 10 & 0 \\ 0 & 0 & 1 \end{Bmatrix} z_k \approx \begin{Bmatrix} 100 & 0 & 0 \\ 0 & 10 & 0 \\ 0 & 0 & 1 \end{Bmatrix} \begin{Bmatrix} 1.1 & 1.1 \\ -2.2 & -2.8 \\ 0 & 0.5 \end{Bmatrix} \hat{x}_k(-) + \varepsilon_k \qquad (4.41)$$

or, applying standard notation used for this problem

$$R^{-1}z_k = R^{-1}H_k\hat{x}_k(-) + \varepsilon_k \qquad (4.42)$$

where it can be seen that the matrix R represents the original mean-squared error noise covariance as assigned to the relative accuracy of the measurements:

$$R = \begin{Bmatrix} 1 & 0 & 0 \\ 0 & 10 & 0 \\ 0 & 0 & 100 \end{Bmatrix} \frac{1}{100} \qquad (4.43)$$

Provided that R has an inverse (it will not if, for example, the pilot thinks he is perfect and thus causes a zero to appear on the diagonal of R), the matrix equation (4.42) is solvable. The conventional solution is to solve Eq. (4.42) for $\varepsilon^T\varepsilon$, take the derivative with respect to $\hat{x}_k$, and set it equal to zero. The resulting expression is the state estimate after update $\hat{x}_k(+)$ and is given by

$$\left(H_k^T R^{-1} H_k\right)^{-1} H_k^T R^{-1} z_k = \hat{x}_k(+) \qquad (4.44a)$$

If the equation set is unweighted ($R = I$), then

$$\left(H_k^T H_k\right)^{-1} H_k^T z_k = \hat{x}_k(+) \qquad (4.44b)$$

Once again the plus sign is used here to indicate that an *updated* estimate of position errors exists at time t_k based both on a priori knowledge and on the set of measurements.

The preceding solution has been programmed in many forms for many years on a wide variety of computers. It will always give a solution as long as there are at least two independent measurements, and there are no limits to the number of measurements or in what order they are admitted. Note that both the latitude error and the tilt error receive an updated estimate even though only one of the states, latitude, was measured.

An estimate can also be made as to what the new accuracy is for the mean-squared value of $\hat{x}_k(+)$ after the update. A reasonable prediction for the covariance of the updated state errors is

$$\left(H_k^T R^{-1} H_k\right)^{-1} = P_{\hat{x}}(+) \tag{4.45}$$

which is the covariance matrix for the error estimates after update, also referred to as the inverse of the information matrix. Thus the more information there is, the less uncertainty.

As will be shown in Chapter 6 for GPS, the square root of the trace of the inverse of the information matrix with $R = I$ is a very important metric called the geometric dilution of precision (GDOP). Increasing GDOP values indicate that error estimates are becoming more uncertain. These values can be used to flag the user that something is amiss or that the information should be used with caution. As applied to GPS, it will be shown that GDOP is heavily influenced by relative satellite geometry.

Linear algebra has very interesting insights involving mathematical projections based on the equations as presented here. This math is the basis for Rudolf E. Kalman's work that has led to the near universal use of the Kalman filter, which in effect is an iterative solution to a weighted least-squares problem between and at the times of update. This important concept is developed in the next chapter, where it will be shown how the weighted least-squares solution to Eq. (4.44) can be solved recursively.

4.4 Closure

One of the most important applications of the weighted least-squares solution was in providing midcourse navigation guidance to the Apollo space vehicle that first carried astronauts to the moon and back. Dr. Harry Goett, director of NASA Goddard Flight Center, initiated work on this problem in 1959 by recruiting the resources of NASA Goddard, NASA Ames, and NASA Langley Flight Centers. Dr. Clarence Gates at the Jet Propulsion Lab (JPL) developed the lunar trajectories, and the linear perturbation technique allowed the least-squares solution to be applied in determining the navigation state estimates.

However, the iterative weighted least-squares estimator in use at JPL required computers that were too large to be put on the spacecraft. A new and advanced estimator had been developed by Norbert Wiener at Princeton, but no one understood how to implement his theory to the spacecraft guidance problem. It was in 1960 that Dr. Stan Schmidt at NASA Ames realized that Dr. Rudolf Kalman, who had developed a linear estimator for state-space systems, had the answer to the lunar guidance problem suitable for small digital computers. The resulting work at NASA Ames was publicized, and it strongly influenced the algorithm that was eventually developed for the small, on-board Apollo computer under the supervision of Battin and Potter at the Massachusetts Institute of Technology. The work also became the basis for a revolution in estimation theory, the Kalman filter, which is the subject of the next chapter.

5
Kalman Filter Inertial Navigation System Flight Applications

A Kalman filter is a computer algorithm—a very famous one. Navigation outputs such as position, velocity, and attitude (PVA) must be updated in order to be bounded. Kalman filtering provides the manner and the method for combining updates that is practically useful as well as mathematically ingenious. Moreover, this technique can be extended to the integration of outputs from a wide variety of systems and continues to show a high degree of practical utility in flight applications.

This introduction presents the rationale for the development and rapid implementation of the Kalman filter in aerospace. It is essential that the previous chapter on uncertainty in navigation systems be completed before proceeding. Some history is presented at the beginning of the chapter, and a short background on Rudy Kalman is in the closure.

The Kalman filter is a linear filter. Recall that the actual differential equations for INS operation are nonlinear, but the error equations are valid for linearized versions of these differential equations. Hence the requirement for the errors themselves is to remain small, otherwise a linear analysis is not valid. Kalman filter applications presume that state-space dynamic modeling will be used to implement the algorithm. In the next part of this chapter, the discrete Kalman filter in its most universal form will be presented and derived in a straightforward manner. Only those parts of the derivation will be emphasized that have practical significance for flight applications.

After you finish this chapter, you should be able to do the following:

1) Describe the practical significance to flight navigation of linear, stochastic models in describing position, velocity, and attitude error propagation under conditions of uncertainty.

2) Identify the well-accepted symbols and common notation used to mathematically describe linear error propagation in inertial guidance testing and evaluation; relate these concepts to nonlinear observations made in the cockpit during flight.

3) Explain the significance of the Kalman filter properties of recursion, optimality, and tuning (including measurement rejection criteria) and introduce the effects of nonlinearities (Extended Kalman filter); state clearly how the Kalman filter gains balance propagation uncertainty with measurement accuracy to produce a "most probable" estimate of position, velocity, and attitude as displayed to the pilot.

4) List, explain, and interpret the causes and effects of Kalman filter divergence as most often evidenced in flight.

5) Explain how an INS aided by GPS can mutually eliminate the major source of errors in each while enhancing both reliability and integrity of the combined

96   INTEGRATED NAVIGATION AND GUIDANCE SYSTEMS

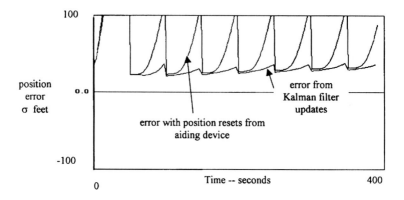

Fig. 5.1   Aided INS performance.

system. The improvement in error estimation using Kalman filter updates over simple position resets is depicted for a typical navigation system in Fig. 5.1.

5.1   Preliminaries

5.1.1   Historical Background for the Kalman Filter

The first widely known important Kalman filter application was for the Apollo moon flight. The midcourse navigation correction problem had its roots at NASA Ames in the search for an algorithm that could be carried on a computer and guide the spacecraft about a known, nominal trajectory. Handling the navigation data that flowed from space and using least-squares fitting techniques taxed even the best computer of that time, and it appeared hopeless to search for an approach that would fit in a small, on-board system.

When Stanley F. Schmidt of NASA Ames Research Center heard Rudolf E. Kalman present his now famous paper at NASA Ames in 1960, he realized immediately that a practical answer was at hand for the space guidance problem. Kalman's linear filter theory seemed to blend perfectly with the trajectory perturbation approach used at the Jet Propulsion Laboratory that resulted in linearized equations of motion. These equations were valid if the corrections were small. The resulting algorithm operated only on the "errors" between estimated and reference trajectories. Simulations confirmed the validity of the approach.

The effort was so successful that a new type of Kalman filter emerged from the work, one which was designed to iteratively linearize about the current state estimate while applying the algorithm—this became the extended Kalman filter. In space application the extended Kalman filter converged even in the presence of significant nonlinearities (although large overshoots tended to occur during early measurement processing).

The 1961 work achieved a potentially significant result for on-board navigation systems. Studies indicated that the extended Kalman filter would be competitive with the weighted least-square estimator with a reduction in computer memory and computation speed. New techniques were developed to process data backward as well as forward, to learn and apply the new state-space approach to dynamics modeling, and to develop data compression methods.

KALMAN FILTER INERTIAL NAVIGATION SYSTEM

In 1966 the Kalman filter was applied to the new C5A aircraft navigation system, and this highlighted a major application problem. The covariance matrix containing the variance of errors must have positive values on the diagonal because they are squared values. Error propagation formulas, however, often combined with other errors to generate an "impossible" covariance that caused the Kalman filter to diverge in flight applications. The answer to this problem was to propagate the covariance in a square-root form, so that during computation the variances would be reasonable (although not necessarily accurate).

5.1.2 Early Kalman Filter Experiences with Radionavigation Aids

In the 1970s an initial effort was made to automatically provide TACAN updates in flight to an inertial navigation system. It was found that real measurements seldom behave as desired. A TACAN receiver that is operating perfectly can still have good quality data interspaced with bad and "wild outliers." In fact, fully useable radionavigation aids often have a few good data points interspaced with large amounts of useless measurements. The development of data rejection algorithms for taking care of the bad measurements that are not flagged by the ground source remains a system integrity problem to this day.

5.2 Kalman Filter: Discrete Case Derivation

This would be a good time to review Sec. 4.1.2.3 on the state-space formulation of error models for inertial navigation systems. There it was emphasized that the navigation error propagation equations for an INS were well known. If initial estimates of navigation error sources were used to drive the navigation error equations on board the aircraft, all nine positions, velocity, and attitude navigation errors would be known and could be used to correct computed INS navigation outputs before they were displayed in the cockpit. If accurate, these navigation error estimates could be used to drive the INS platform toward a correct alignment using a feedback scheme, but this was concluded to be very risky in flight. The problem with this line of thinking is that the guess for the initial navigation errors would always be both inaccurate and incomplete. Because such a wide variety of error sources exist in an inertial navigation system, the propagated equations if unaided would be increasingly inaccurate as time passed.

5.2.1 Motivation for the Kalman Filter for Inertial Navigation System

These are the major problems that exist if one relies on an inertial navigation system for air navigation: how to correct the navigation error equations while flying so that they remain useful even though the initial navigation errors were not known accurately; how to deal with noisy measurements from a variety of other systems that are arriving at different times; how to estimate the covariance of the INS output whenever an update occurs to see how much of the measurement should be believed in the presence of noisy system dynamics; and finally, how to obtain estimates for all navigation outputs even though only one or two is being measured by other means, providing as a result the most probable position (MPP).

The Kalman filter can be used to solve all of these problems. The basic procedure goes like this: Initialize the filter by providing statistical estimates $\hat{x}_o$ for the initial navigation error states, their covariance $P(t_0) = P_o$, and noise sources Q_o.

Run the computer to propagate, using Eq. (4.37), both $P(t_0)$ and the $\hat{x}_o$ to time of measurement update t_k before the measurement z_k. Note that values before measurement are $P_k(-)$ and $\hat{x}_k(-)$. Compute the most probable estimate $\hat{x}_k(+)$ by weighting $\hat{x}_k(-)$ and z_k with the Kalman gains (as discussed in the following).

If the filter converges, then you have a useable estimate of all of the navigation error states at all discrete times t_k during the flight. You can propagate the navigation error states to any time and combine them with measurements from any source no matter how noisy. You also have an estimate of the covariance of the navigation error states and can verify that it decreases following a measurement update. Finally, you have estimates for *all of the navigation error states* even if only one of them is measured. This is a powerful tool—when it converges.

The catch of course is to weight $\hat{x}_k(-)$ and z_k in just the right way. This is done using what is now universally called "the Kalman gain matrix" K. Another possible catch is knowing when to accept and when to reject the measurement updates. Finally, it is difficult to get accurate estimates for the noise covariance associated with the error sources and with the measurements. Establishing the right balance in choosing Q_k and the measurement error covariance R_k is difficult and has been called *tuning* the Kalman filter.

5.2.1.1 Kalman gain determination.
The discrete Kalman filter provides an iterative solution to the least squares estimate from Eq. (4.44)

$$\hat{x}_{k_{(\text{LeastSq})}} = [H^T R^{-1} H]^{-1} H^T R^{-1} z_k \tag{5.1}$$

by implementing a solution for the system described in Table 5.1.

From Table 5.1 form the navigation error residual for the navigation error *before* the measurement update as follows:

$$\tilde{x}_k(-) = \hat{x}_k(-) - x_k \tag{5.2}$$

Form the navigation error residual for the navigation error *after* the measurement update:

$$\tilde{x}_k(+) = \hat{x}_k(+) - x_k \tag{5.3}$$

From Table 5.1

$$\hat{x}_k(+) = K'_k \hat{x}_k(-) + K_k(H_k x_k + v_k) \tag{5.4}$$

From Eq. (5.3)

$$\tilde{x}_k(+) = K'_k \hat{x}_k(-) + (K_k H_k - I) x_k + K_k v_k \tag{5.5}$$

Use Eq. (5.2) to eliminate $\hat{x}_k(-)$ and get

$$\tilde{x}_k(+) = (K'_k + K_k H_k - I) x_k + K'_k \tilde{x}_k(-) + K_k v_k \tag{5.6}$$

It is desired to have the expected value of each of these terms equal to zero because this is a navigation error residual, and so this implies

$$K'_k = (I - K_k H_k) \tag{5.7}$$

Substituting into Table 5.1,

$$\hat{x}_k(+) = (I - K_k H_k) \hat{x}_k(-) + K_k z_k \tag{5.8}$$

$$\hat{x}_k(+) = \hat{x}_k(-) + K_k [z_k - H_k \hat{x}_k(-)] \tag{5.9}$$

KALMAN FILTER INERTIAL NAVIGATION SYSTEM

Table 5.1 Discrete state-space system

Equation Description	Equation
Discrete system equation and error source	$x_{k+1} = A_k x_k + w_k$
Noise covariance of error source noise	$E(w_k w_k^T) = Q_k$
Discrete measurement equation	$z_k = H_k x_k + v_k$
Noise covariance of measurement	$E(v_k v_k^T) = R_k$
Assumed form for estimate of navigation error state (must find K_k' and K_k)	$\hat{x}_k(+) = K_k' \hat{x}_k(-) + K_k z_k$ $K_k' = (I - K_k H_k)$
Final form for estimate	$\hat{x}_k(+) = \hat{x}_k(-) + K_k[z_k - H_k \hat{x}_k(-)]$
[from Eq. (5.19)]	$K_k = P_k(-) H_k^T \left[H_k P_k(-) H_k^T + R_k \right]^{-1}$
Covariance of navigation error residual $P_k(+) = E[\tilde{x}_k(+) \tilde{x}_k^T(+)]$	$P_k(+) = (I - K_k H_k) P_k(-)$ $P_k^{-1}(+) = P_k^{-1}(-) + H_k^T R_k^{-1} H_k$
Propagation equations for estimate $\hat{x}_{k+1}(-)$ before update z_{k+1} and for its residual covariance $P_{k+1}(-)$	$\hat{x}_{k+1}(-) = A_k \hat{x}_k(+)$ $P_{k+1}(-) = A_k P_k(+) A_k^T + Q_k$ $Q_k = E(w_k w_k^T)$
Alternate form for K_k	$K_k = P_k(+) H_k^T R_k^{-1}$

The estimated measurement before the update is

$$\hat{z}_k(-) = H_k \hat{x}_k(-) \tag{5.10}$$

The measurement residual, sometimes called the "innovations," is given by

$$\tilde{z}_k = z_k - \hat{z}_k(-) = H x_k + v_k - H \hat{x}_k(-)$$
$$= H \tilde{x}_k + v_k \tag{5.11}$$

The Kalman gains are found starting with Eqs. (5.6) and (5.7)

$$\tilde{x}_k(+) = K_k' \tilde{x}_k(-) + K_k v_k$$
$$= (I - K_k H_k) \tilde{x}_k(-) + K_k v_k \tag{5.12}$$

From Eq. (5.2) and (5.12) the covariance of the error residual $\tilde{x}_k(+)$ is

$$P_k(+) = E[\tilde{x}_k(+) \tilde{x}_k^T(+)]$$
$$= E\{[(I - K_k H_k) \tilde{x}_k(-) + K_k v_k][(I - K_k H_k) \tilde{x}_k(-) + K_k v_k]^T\}$$
$$= (I - K_k H_k) E\{\tilde{x}_k(-) \tilde{x}_k^T(-)\} (I - K_k H_k)^T + K_k E\{v_k v_k^T\} K_k^T$$
$$+ (I - K_k H_k) E\{\tilde{x}_k(-) v_k^T\} K_k^T + K_k E\{v_k \tilde{x}_k^T(-)\} (I - K_k H_k)^T$$
$$\tag{5.13}$$

The covariance $P_k(-)$ before the measurement update and the covariance of the measurement noise are

$$P_k(-) = E[\tilde{x}_k(-)\tilde{x}_k^T(-)] = P_k^T(-) \quad (5.14a)$$

$$R_k = E[v_k v_k^T] = R_k^T \quad (5.14b)$$

and assuming that measurement errors are uncorrelated with error source noise

$$E[\tilde{x}_k(-)v_k^T(-)] = E[v_k(-)\tilde{x}_k^T(-)] = 0 \quad (5.15)$$

Equations (5.15) and (5.14) may be substituted into Eq. (5.13) to yield

$$P_k(+) = P_k(-) - [P_k(-)H_k^T K_k^T] - K_k H_k P_k(-)$$
$$+ K_k H_k P_k(-) H_k^T K_k^T + K_k R_k K_k^T$$
$$= P_k(-) - P_k(-) H_k^T [H_k P_k(-) H_k^T + R_k]^{-1} H_k P_k(-) \quad (5.16)$$

The Kalman gains K_k will be chosen to minimize the sum of the diagonal (trace) of $P_k(+) = E[\tilde{x}_k(+)\tilde{x}_k^T(+)]$, which will result in the most probable state estimates for $\hat{x}_k(+)$. To do this, a result from linear algebra must be applied that states

$$\frac{\partial}{\partial K_k}\{tr\, K_k R_k K_k^T\} = 2K_k R_k \quad (5.17a)$$

$$\frac{\partial}{\partial K_k}\{tr\, A K_k^T\} = A \quad (5.17b)$$

$$\frac{\partial}{\partial K_k}\{tr\, K_k A\} = A^T \quad (5.17c)$$

After applying Eq. (5.17) to Eq. (5.16),

$$\frac{\partial}{\partial K_k}\{tr\, P_k(+)\} = -2P_k(-)H_k^T + K_k H_k P_k(-) H_k^T + K_k R_k K_k^T \quad (5.18)$$

Setting Eq. (5.18) equal to zero and solving for K_k provides

$$K_k = P_k(-) H_k^T [H_k P_k(-) H_k^T + R_k]^{-1} \quad (5.19)$$

which can be back substituted into Eq. (5.16) to provide

$$P_k(+) = (I - K_k H_k) P_k(-) \quad (5.20)$$

Finally, the matrix inversion lemma from linear algebra may be applied to the second part of Eq. (5.16) to yield

$$P_k^{-1}(+) = P_k^{-1}(-) + H_k^T R_k^{-1} H_k \quad (5.21)$$

Note that this can be verified by multiplying Eqs. (5.20) and (5.21). Multiplying the three Eqs. (5.19–5.21) yields an alternate form for

$$K_k = P_k(+) H_k^T R_k^{-1} \quad (5.22)$$

The last two equations give an alternate method of solving for the covariance of the navigation error residual and updating the Kalman gains. The full order inverse in Eq. (5.21), however, makes this form impractical. The inverse in Eq. (5.19) has a much lower order equal to the dimension of the measurement error covariance matrix R_k.

Schmidt (1995) and Strang (1997) show in detail how the Kalman filter solution for the state estimate that is given in Table 5.1 matches the weighted least squares solution in Eq. (5.1). The key is to weight the measurements by the inverse of their covariances.

5.2.1.2 Covariance propagation equation. With the value calculated for the Kalman gain matrix K_k, the new state estimate $\hat{x}_k(+)$ after update using measurement z_k is propagated (computed) to the next measurement time (see the state-space formula in Table 5.1). At the next measurement time the propagated estimate of the navigation error state is $\hat{x}_{k+1}(-)$ where the minus indicates that the estimate is before the measurement update using z_{k+1}. The covariance of the estimate is propagated also using Eq. (4.37)

$$P_{k+1}(-) = A_k P_k(+) A_k^T + Q_k$$

$$Q_k = E(w_k w_k^T)$$

[see Sec. (4.2.3.4)] where Eq. (4.37) was derived in the previous chapter on uncertainty. The Kalman gain matrix is then computed for the new time K_{k+1} and so on.

Congratulations if you made it this far. These equations are nontrivial and require much experience to work with successfully. One of the interesting things to note about the navigation error residual covariance can be seen from Eq. (5.21). The formula is similar to resistors in parallel, and from circuit theory we know that the net resistance must be less than any of the resistors. This implies that the updated navigation error residual covariance is less than either the updated navigation error residual covariance or the measurement error covariance. This sounds too good to be true, but it can have nasty implications in practice, leading to divergence of the filter. This is the topic that follows.

5.3 Kalman Filter Divergence

This is an important topic. Pilots are often given the capability in the cockpit to accept or reject a measurement update, and making a wrong decision can lead to *loss of situational awareness and assessment (SA/A)* at very critical times. One of the most important "pilot-in-the-loop" questions is how much authority and capability should the pilot be given by the engineering design team in operating complex integrated avionics systems. Certainly the pilot should not have the capability to tune the Kalman filter gains, but neither should the pilot be prevented from entering correct updates to the navigation system. The discussion that follows is intended to shed some light on this area and to provide insight on system performance and integrity, but it will not provide any "cookbook" answers. The pilot, as always, remains responsible for SA/A.

Kalman filter divergence has too many forms to predict what it will look like in a modern cockpit, but suffice it to say it is a bad thing to occur. Bad measurements may be accepted, good measurements rejected, and the navigation computer may have no source to tell it things have gone wrong. This is a loss of what is called *system integrity*. As the pace of system integration accelerates, especially in GPS-aiding INS (and vice versa) applications, these problems become increasingly important.

5.3.1 Computer Word Length

The problems of insufficient computer speed and word length (how accurate numbers can be represented in a computer) were obvious in early Kalman filters, and much work was expended on fixed-point machines that operated using integer representations. Modern floating point systems may mask this problem in the following way: a 32-bit fixed-point computer provides a greater accuracy than a floating point computer with 24-bit mantissa and 8-bit exponent, but both are referred to as 32-bit computers. Kalman filters in aerospace are designed to operate in real time, and few engineers are trained to program software for these systems when the integer bit size must be carefully taken into consideration.

5.3.2 Monte Carlo Simulations

Monte Carlo simulations of navigation performance directly address otherwise intractable issues in design, development, and flight testing. An INS, for example, has a wide variety of error sources, each of which may have random variables at system startup and during flight. Some of these sources may cause divergence consistently or only occasionally. If a Monte Carlo simulation is performed on the INS, error source inputs for all runs are statistically selected, and a large number of runs are performed so that statistical averaging can occur across event space (instead of time averaging for one sample). The simulations can be modified to select some error sources as deterministic and some as random (the same technique can be used with environmental variables that affect system performance).

Families of error plots result from Monte Carlo simulations from which reasonable evaluations of system performance and capability may be predicted. Moreover, the simulations may expose combinations of error sources that are extremely detrimental in the flight environment so that these issues can be addressed and suitably modified. Monte Carlo simulations are used extensively in investigating Kalman filter divergence problems and solutions. This is especially a good idea before operational flight program (OFP) software changes or software integration. Monte Carlo simulations will be demonstrated in Sec. 10.3.1.

5.3.3 Nonlinear System Behavior

Remember that the Kalman filter is a linear filter in that K_k, the Kalman gain matrix, produces a weighted linear combination of old estimates and measurements to produce new estimates. The Kalman filter, however, operates in a very nonlinear world. The aerospace community has stretched the original filter to operate on nonlinear, continuous system dynamics with discrete time measurements. The result is called the extended Kalman filter (EKF). In the EKF the nonlinear system is linearized "on the fly" about the propagated state estimate (generating a new, linear state-space system), and the measurements are linearized about the estimated measurement (generating a new measurement matrix H_k). The real-world measurement is then applied to these newly linearized models, and so the Kalman gains K_k must be computed every time (and not pre computed and stored in the computer as in some applications). The computer burden of the EKF is high. The linearization process will be demonstrated in Chapter 6.

Recall that the perturbed error equations for an INS were linear, so a new state-space system did not have to be generated at each measurement update time. However, the update measurements applied to these equations may be nonlinear, requiring that a new H_k be obtained at each update time t_k (this is the case if GPS measurements are used). The bottom line here is that the EKF has lost a lot of the optimality associated with Kalman filtering, and all of the guaranteed stability (Kalman filtering is theoretically stable when applied to observable linear systems). The more dynamic the maneuvering task, the harder the EKF has to work, and probably the more inappropriate it is for the mathematics that describe the scenario. This is not to say that the EKF will not converge during high dynamics situations, but that it will only converge consistently and well in the specific maneuvering scenario for which it was tested and designed. Expect divergence of the EKF for other applications.

5.3.4 Covariance Matrix Calculations

The problem of Kalman filter (KF) divergence caused by covariance calculations, which has been with system developers since the 1960s, is still with the aerospace community. Basically, the covariance matrix $P_k(+)$ becomes too small resulting in a small $K_k = P_k(+)H_k^T R_k^{-1}$ and thus eliminates the weighting on new measurements as can be seen (from Table 5.1)

$$\hat{x}_k(+) = \hat{x}_k(-) + K_k[z_k - H_k x_k(-)] \qquad (5.23)$$

This results in the new state estimate after update staying the same as the estimate before update (because K_k approaches zero). This estimate is then propagated forward, and the same thing occurs at the next measurement time. The filter, in effect, is rejecting all of the new measurements and relying only on its own propagated values for covariance which keep getting smaller and making things worse. This is a loss of *system integrity*.

Believing in itself too much can be considered a case of "Kalman filter incest," and it is commonly addressed by injecting a hefty dose of computer generated covariance into $P_k(+)$ at update time so that the filter will continue to weigh the measurements according to $K_k = P_k(+)H_k^T R_k^{-1}$. Of course this amounts to little more that setting our own, relatively arbitrary gains, and can easily allow bad measurements to influence the gain calculations. This has adverse problems of its own, discussed later.

The underlying problem is that the models are not accurate enough or that inappropriate statistics have been input into the filter algorithm. Kalman filter incest is especially difficult to flag because the state-space propagation equations provide smooth, uninterrupted solutions that are being continually sent to the cockpit and to other avionics. When the pilot looks out the window and discovers that a large update is required, a serious problem may result when the update is sent to the computer. Either the computer will consistently reject the input as a bad measurement, or the computer, if forced to "swallow" it, may correct the state estimates too far and too fast to keep the filter algorithm stable. Remember that the KF is a linear algorithm and errors must be kept small in linear systems if they are to describe real world behavior.

5.3.5 Inaccurate Measurement Models

Measurements typically have bias and scale factor errors that must be taken into account when updating. The problem is that they may change from time to time, just like the driving error sources like gyro bias in an INS. The Kalman filter takes a unique and ingenious approach to this problem. It presumes a model form for these phenomena and adds states to the state-space model that, when estimated, will provide reasonable values for biases and scale factors if the KF converges.

This rather amazing property of being able to estimate error sources and measurement biases on the fly results because the Kalman filter is, in the vocabulary of control and estimation theory, a "mathematical observer." This is not just a by-product of the Kalman filter technique—the observer properties were mathematically built into the filter from the beginning. It is one reason that the number of states in a Kalman filter implementation tend to get out of hand. The INS only has nine navigation error states, but by the time all the source and measurement errors are modeled there may be over 50 states active in the filter. This is called *augmenting the state-space system*.

Results from Monte Carlo simulations used to tune the filter may indicate that setting an rms to zero in the covariance inputs has little effect on performance. If this corresponded to a modeled measurement bias, for example, then the bias could be eliminated from the state-space model and the order of the equations reduces with little loss of performance. This, of course, is error budgeting, a very important objective of Monte Carlo simulations.

5.3.6 Erratic Ramp/Reset Modes

After a measurement update the estimated state is reset, so to speak. If this estimated state is used for feedback control, its effects will cause a jump in the system dynamics that could be very objectionable. A software change could filter the state into the control system, but this adds lag that will exist at all times, even when the filter is not updating. Lag in a control system is always adverse (possibly to the point of instability). A technique used by Stan Schmidt at Northrop was to reverse the estimated state update in the computer and then put a lag filter on it so that the reversed signal only exists during update. This signal is then added to the actual update so that it enters the control system gradually but only when an update occurs.

5.3.7 Poor Filter Initialization

This topic was introduced in Sec. 5.3.5 and can cause large errors. A standard technique to address this issue is to accept measurements during filter initialization. Then, after discarding the first few that normally will cause a problem, perform a least-squares solution that provides accuracy statistics in the form of covariance. The filter is then restarted with this initial data.

5.3.8 Measurement Acceptance/Rejection Criteria

If "outlier" measurements cannot be rejected, the navigation error will rapidly grow without bound because the filter equations will ignore their own system equations and become erratic. This is the opposite of the Kalman filter incest discussed in Sec. 5.3.4 where the filter ignored the measurements and "marched to its own drum." Measurement rejection criteria are normally based on the innovations given by $[z_k - H_k \hat{x}_k(-)]$ or the equivalent $[z_k - \hat{z}_k(-)]$, which represents the new information that the measurement z_k is providing.

If measurements are processed one at a time, then the denominator of the Kalman gain $\sqrt{[H_k P_k(-)H_k^T + R_k]}$ given by Eq. (5.19) is the rms of the innovations $[z_k - H_k \hat{x}_k(-)]$. Because these quantities are available in the filter, they may be compared. If the magnitude of the innovations is more than three times the rms of the innovations, then a "3σ test" would consider the measurement z_k an outlier and reject it. A human operator is often allowed to override the rejection. During those portions of the flight where a "self test" assures good measurements, a "5σ test" or "6σ test" may be used.

5.4 Aided Inertial Navigation System and the Inertial Navigation System Error Budget

Inertial systems are here to stay. Their passivity and usefulness in a threatening environment are unchallenged. It is how they are used that is changing. In a very real sense the INS and the GPS are now mutually aiding, each covering the other's shortcomings and benefiting the overall system performance.

This discussion still presumes the INS as *primary* and the other sources as *aiding*, that is, the other sources provided measurements for the Kalman filter. This implementation strategy is well understood with many lessons learned to impart to the developers and testers of the next generation of avionics systems. In this strategy it is the job of the INS to keep position, velocity, and attitude (PVA) current, and the job of external measurements to enter the Kalman filter and thus to produce what we have called the *most probable position* (MPP) in the overall navigation solution.

5.4.1 Kalman Filter Computer Location

The aircraft avionics must define the interfaces between the navigation system and other systems. The inertial measurement unit (IMU) contains the sensor packages and the computer to solve the Fundamental Equation of Navigation (FEN) thus to produce the estimated (computed) state vector. The EKF combines the state estimates with raw measurement data from external sources and produces a correction to the state estimates as well as a covariance matrix that tells how accurate the corrections are.

The Kalman filter computations are not required to be in the IMU computer. They only require access to the navigation solution. Where and how to perform the filtering correction is not standardized and has been the subject of much discussion. The solution has been mission specific in the past, but increasing cost emphasis has led a drive to standardize the implementation. The important point to remember is that the Kalman filter works best when it is working with raw data. If incoming data has already been processed by other Kalman filters in other units, then estimates may be correlated with the measurements and lead to erratic performance. N. A. Carlson has developed a federated Kalman filter theory and implementation strategy to address this issue, and more work is underway in this area. Special attention and caution is required when the measurements control units are themselves aiding the INS.

5.4.2 Computer Simulation Packages

The way to learn the most from flight-testing aided inertial navigation systems is to have developed prior to flight a set of simulation packages that can be used to duplicate, as much as practical, the actual flight-test results when they occur. Remember that the computer can be a significant source of error, and it would be very desirable to test out the filter—if not with the actual hardware—then with a simulation of the hardware that represents worst case situations. The arithmetic operations of the simulation computer should be compared with those of the flight computer to avoid building an optimist's Kalman filter instead of an optimal one.

To test the actual flight computer routines, a means of simulating the IMU data and the measurement of raw data should be found. Trajectory generation packages are available, which can be used for "truth" data and to provide simulated sensor outputs to a simulated INS that accelerometers and gyros provide in-flight to the real INS. The net result is a *flight systems simulator* for most or all of the key systems that are impacted by the on-board computer program.

5.4.3 Flight Testing the Aided System

This is the final step of the validation and refinement of the aided system. The flight should have enough data taken to not only provide a measure of performance such as CEP, but to understand and to correct anomalies before the system becomes operational. Measure the flight test against this standard: Can the flight-test recorded data be used with the flight simulation system to duplicate the flight results? A validated flight simulation system would then become enormously valuable when doing Monte Carlo runs that would cost a small fortune in a flight-test program.

5.4.4 Developing the Error Budget

The objectives of an error budget are to determine the qualitative relationships between sensor accuracy, the number of sensors, the accuracy and type of measurements from potential aiding sources, the environmental constraints, and overall navigation system performance. Error budgets for the software include the number of states for the filter, for modeling the error sources and measurements, and even for the type and size of computer. This analysis is in reality a tradeoff study with cost, performance, and operational need in the driver's seat. The outputs of an error budget are the sensor and system requirements needed for a specified set of operational capabilities, and if possible a sensitivity study highlighting the most important system integration tradeoffs.

5.4.5 Performance of an Aided Inertial Navigation System

The type of performance one might expect from an aided INS system is shown in Fig. 5.1 where the familiar "sawtooth" performance plot is easily seen. The start of the Schuler cycle can be seen at each update measurement time. The most important thing to note is that the rate of change of the error is much slower if a Kalman filter operates on the measurements. Note also that without the Kalman filter the error at a performance update can never be better than the measurement. The Kalman filter, however, as was derived and explained, produces a covariance that is less than the covariance of either the propagated state estimate or the measurement.

5.4.6 Global Positioning System—Aided Inertial Navigation System

The following are both summary paragraphs for this chapter and an introduction to the next part of the text (GPS-aided INS cannot be fully treated until the chapters that describe GPS are completed, but the topic is of such importance that it is introduced here). The GPS will be shown to be a position-velocity-time (PVT) device; the INS is a position-velocity-attitude (PVA) device. GPS has good, stable long-term performance but usually provides outputs at a slow rate (once per second). The INS has fast, stable short-term performance and provides outputs at a fast rate (15 Hz or more; Kalman filter updating, however, occurs at a much slower rate). The GPS can be used to update the position and velocity INS error states. The INS can help the GPS point to satellites if it loses lock.

Remember that a Kalman filter provides the most probable position when working with raw data. Thus an excellent integration scheme has the INS and GPS both providing raw data to the filter equations. But the primary point to be made is that the combined GPS/INS solution to the navigation problem provides excellent *system integrity* in addition to *exceptional performance*. Thus the *overall system performance* is vastly improved over that of either system acting alone when considering multiple quantitative and qualitative metrics. No matter which system is riding shotgun for the other, together they form a formidable pair.

This completes what may be considered the first part of the text that by necessity had to cover many theoretical and mathematical topics that are taught in graduate-level engineering courses. The remainder of this work consists primarily of applications, except where additional theory must be covered to understand the application itself.

5.5 Closure

Norbert Wiener was a great mathematical researcher at Princeton who studied controllers for antiaircraft fire using noisy radar data in World War II. Wiener's resulting work in estimation theory was classic, and after the war it was published in a report titled *Extrapolation, Interpolation, and Smoothing of Stationary Time Series*. Norbert's students affectionately called the book "The Yellow Peril" because it was written on yellow, legal pad paper. By the early 1950s the emerging field of statistical estimation theory was strongly influenced by Wiener's many contributions. (The discrete-time formulation of Wiener's work was completed overseas by A. N. Kolmogorov in 1941).

Rudolf Emil Kalman was born May 19,1930, in Budapest, Hungary. He immigrated to the United States with his family because of the disruption of the Second World War. The idea of a discrete state-space filter as a solution to Norbert Wiener's estimation problem occurred to him on a train trip to Princeton from Baltimore in 1958. His idea was greeted with skepticism at universities, and one of his now classic papers was rejected with the comment that "it cannot possibly be true." His linkage with Stanley Schmidt at NASA Ames Research Center thus was the key in giving his work the exposure that it was due. The Kalman filter, especially in its nonlinear formulation as the extended Kalman filter, is one of the greatest discoveries in the history of statistical estimation theory.

6
Global Positioning System

The global positioning system (GPS) was initiated in the 1960s as an all-weather navigation system with global coverage, operated by the military, but with a few planned spin-off capabilities for commercial users. After a rocky start, enabling technologies in microelectronics and communications spawned some commercial interest that has rapidly exploded into what has been called one of the major technological innovations of the 20th century. GPS use in the Gulf War in the early 1990s accelerated the already strong technological pace that continues unabated today.

GPS is assured to continue as a revolutionary idea and as a rejuvenating force in the aerospace industry well into the 21st century. It has been proven in the stern test of combat and has spawned a furious competitive development in industry. Low-cost GPS units have already impacted and will soon totally change the air traffic control system. In all services of the military, GPS provides stand-off weapons delivery capabilities, autonomous aircraft, and precise positioning for vehicles. Radionavigation aids to flight will be steadily replaced or combined with GPS. Within the continental U.S. the wide area augmentation system (WAAS), a network of three satellites and ground monitoring stations, will permit precision instrument approaches at nearly any airport or suitable landing spot. In short, GPS is an idea whose time has surely come.

In this and remaining chapters, the GPS will be described in detail with an eye toward flight applications and flight testing. The basic theory needed for a thorough appreciation of GPS and its potential has been presented in previous chapters, and only the added theory needed for understanding a particular application will be presented from this point on. The remaining chapters are not long and aim to give the reader a window into other sources about GPS. This is important because the technology is changing so fast that anything written about GPS or its applications before 1995 already may be severely dated.

After completing this chapter you will be able to do the following:

1) Explain the history, overview, and potential impact of satellite navigation on all navigation systems worldwide, and appreciate the synergism between military and commercial developments.

2) Review the reference coordinate systems for satellite navigation, especially the Earth-centered Earth-fixed (ECEF) system, and relate them to fundamental geodetic properties of the Earth as defined by the WGS-84 ellipsoid.

3) Outline the three major GPS segments (space, ground, user) and list the important characteristics of each; explain the basic concept of using time-of-arrival measurements for position fixing and the basic limiting factors on solution update rates; list and explain the major types of time (UTC, GPS, USNO, TAI).

4) Identify and list the major sources of error in GPS position and velocity measurements, especially the intentional effects of selective availability; infer and

predict the impact of these error sources on the inflight use of GPS; define and explain the significance of the dilution of precision (DOP) units of GPS error.

6.1 Principles of Operation

6.1.1 Background

The idea for satellite systems for position originated in the 1960s when several U.S. government agencies and the military sought a worldwide three-dimensional navigation capability for all vehicles, on the land and in the air, that was unaffected by weather, time of day, or line of sight limitations. The Navy Transit program (1964) was suited for ships, and the Navy was developing a Timation satellite for time transfer related to two-dimensional navigation.

The Air Force had studied the problem and called their solution System 621B, with the innovation of a new type of ranging technique that employed pseudorandom noise (PRN) to transmit satellite data. This idea had much support because these signals (tested at White Sands Missile Range in New Mexico) could be detected at an amazing 0.01 signal to noise ratio. Even more startling, all satellites could broadcast different PRN codes *on the same frequency* (PRN codes have the mathematical property of being "orthogonal"). However, it was not until the Army at Ft. Mommouth, New Jersey, verified that the time of arrival (TOA) technique using PRN codes, now called *pseudoranging*, was the best method to find position that the Air Force methodology was fully accepted.

At the same time, system survivability (not to mention huge cost) was an issue with Congress. The Department of Defense "circled the wagons" and established the Defense Navigation Satellite System (DNSS) to combine service efforts into a joint-use system. David Packard, of Hewlett-Packard fame, had been pushing joint programs, and GPS was an ideal candidate. Thus the Joint Program Office (JPO) was formed. The program was called NAVSTAR because a key reviewing official thought the name "sounded nice." (A high-ranking military officer had originally suggested the "Global Positioning System" name in 1970 because it had a better sound than mere "navigation.")

In 1973 Col. Parkinson and his JPO team were turned down for a program go-ahead, but in the process won the support of Malcolm Currie, who was a powerful force in defense procurement at the time. Following Currie's advice, Parkinson proposed a synthesis of competing systems and the previously Air Force-only program became truly multiservice. Approval to proceed was granted in December of 1973, and the program was officially renamed the NAVSTAR/Global Positioning System.

6.1.2 Global Positioning System—The System

GPS consists of three segments: space, control, and user. The space segment (see Fig. 6.1) consists of 24 operational satellites in six circular orbits 20,200 km (10,900 n miles) above the Earth at an inclination angle of 55 deg with a 12-h period. The satellites are spaced in orbit so that at any time a minimum of six satellites will be in view to users anywhere in the world. The satellites continuously broadcast position and GPS time data.

The control segment consists of a master control station in Colorado Springs, Colorado, with five monitor stations and three ground antennas located throughout

GLOBAL POSITIONING SYSTEM

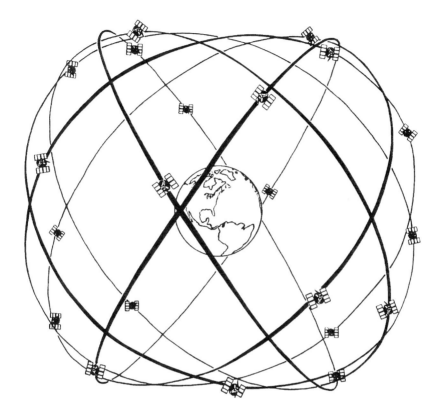

Fig. 6.1 GPS—the system.

the world. The monitor stations track all GPS satellites in view and collect ranging information from the satellite broadcasts. The monitor stations send the information they collect from each of the satellites back to the master control station, which computes extremely precise satellite orbits. The information is then formatted into updated navigation messages for each satellite. The updated information is transmitted to each satellite via the ground antennas, which also transmit and receive satellite control and monitoring signals.

The user segment consists of the receivers, processors, and antennas that allow users to receive the GPS satellite broadcasts and compute their precise position, velocity, and time. The user's receiver measures the time delay for the signal to reach the receiver, which is the direct measure of the apparent range to the satellite. Measurements collected simultaneously from four satellites are processed to solve for position, velocity, and time.

6.1.3 Navigation Accuracy Using Time of Arrival

The typical expected accuracy of the GPS system at moderate latitudes is given in Table 6.1. The HDOP and VDOP are horizontal dilution of precision and vertical dilution of precision values, respectively, and indicate the satellite geometry

Table 6.1 Global positioning system expected accuracy

	Horizontal position, m (one σ: HDOP = 2)	Vertical position, m (one σ: VDOP = 2)
PPS	10	13
SPS	40	50

effects on solution accuracy. Accuracy degrades as the dilution of precision (DOP) increases. The advertised position dilution of precision (PDOP) is 6 or less. The actual average of GDOP is nearer to 2.5. PPS is the precise positioning service available to military users only, while standard positioning service (SPS) coverage is available to everyone. SPS service is intentionally degraded by the military using a concept called selective availability (SA).

The advertised accuracy from the Federal Radionavigation Plan (FRP) is 22 m horizontal accuracy, 27.7 m vertical accuracy, and 100 ns time accuracy for the PPS and 100 m horizontal accuracy, 156 m vertical accuracy, and 340 ns time accuracy for the SPS. Note that the vertical accuracy is worse than the horizontal accuracy. (This difference is caused by the positions of the satellites in the sky relative to the users and by the atmospheric effects of the stratosphere and troposphere.)

6.1.3.1 User positioning from time of arrival.

The excellent accuracy of GPS is due to the ranging technique called time of arrival (TOA). Each satellite has an atomic clock set to GPS time, and GPS time increments are normally slaved to within 20 ns of Greenwich Mean Time (GMT). GMT time, also called Coordinated Universal Time (UTC) or "Zulu" time, is a weighted average from laboratory clocks worldwide and is not continuous but leaps by a second whenever the irregular motion of the Earth demands it. Time kept by the GPS master ground station, however, ignores these leaps and so has built up a time bias of approximately 11 s. (Refer back to Sec. 2.1.3 for a complete description of worldwide time standards.)

Because all clocks slow down with relative motion, synchronization of time among satellites is obtained by hypothetically defining each satellite clock as a nonmoving "snap-shot" location within the inertial C_i frame. (The speed of light is constant in a nonrotating frame.) A relativistic correction to each actual satellite clock is then applied so that its time agrees with the hypothetical clock in the C_i frame. Actual Earth-referenced navigation occurs, then, using the time of the hypothetical set of satellite clocks. This is another way to define *coordinate time* as is done in defining UTC.

Clocks fixed to Earth's ellipsoid at mean sea level (MSL) on the geoid all rotate at Earth rate and keep the same relative time, but this time requires a correction to match coordinate UTC. If a government laboratory is not at the geoid MSL, an additional relativistic correction is required. Thus the U.S. frequency standard runs fast by about 15 ns per day with respect to a geoid clock at MSL. These corrections for GPS satellites amount to approximately 40 ns per day, and compensation for this relativistic effect is applied to each GPS satellite before launch. This allows a user fixed to the Earth to receive the advertised broadcast frequency. Time is transferred between laboratories very accurately by transmitting time differences (GPS *common view* technique).

Corrections to GPS satellite time and the satellite's orbit in C_{ECEF} coordinates are encoded in the transmitted navigation message. This message is received by the user in the navigation message and stored as a computer word. If the user can now determine the GPS time that the message was received, then by knowing the speed of light the distance to the satellite can be computed. The received navigation message occurs in what is called a *measurement epoch*.

Unfortunately, although the user has a very accurate clock (a crystal one) in his GPS receiver, the clock is neither an atomic one nor does it keep GPS time exactly. A bias thus exists that must be determined before the user can compute his position. The TOA that the user has stored in his computer is called the pseudorange. The user also has stored in his computer the position of the satellite at the time the message was sent, which was obtained by using *ephemeris* data contained in the navigation message from each satellite. When the user has done all that he can with the data from one satellite, he knows that his position is somewhere on a sphere centered at the known satellite position with a radius determined by the TOA that is not known exactly due to the time bias.

Recall from Sec. 6.1.1 that these GPS satellites have the amazing capability to transmit on the same frequency at the same time using orthogonal *pseudorandom* (PRN) codes. All satellites thus transmit their message in the same measurement epoch every millisecond. If data from a second satellite is processed in the same measurement epoch as data from the first, the user will know that his position is somewhere on the intersection of two spheres, each centered at the respective satellites and each with an uncertain radius (but uncertain by the same amount because there is only one time bias). A third satellite will provide a solution at the intersection of three spheres, but again with uncertain radius.

From the basic theory of simultaneous equations, we know that four satellites must be used to determine the three components of position and also solve for the time bias. The positioning problem will be illustrated here in two dimensions as shown in Fig. 6.2. The pseudorange measurement from a satellite is equivalent to knowing a "user position" ring about that satellite. If only satellites SV1 and SV2 are considered, note that there is a large position ambiguity at the intersection of the two respective position rings. The user must either know which position is correct or use a third satellite, SV3, to resolve the large position ambiguity.

A lesser but still significant ambiguity exists in the two-dimensional positioning solution. Note in Fig. 6.2 that an unknown time bias will cause the outer radius of the three positioning rings from SV1, SV2, and SV3 to intersect at points **A**, **B**, and **C**, defining a spherical triangle **ABC**. The user receiver calculates the receiver time bias in such a way that spherical triangle **ABC** is minimized, resulting in the least ambiguity in a least-squares sense for a position solution in two dimensions. If a fourth satellite is added, a three-dimensional solution can be calculated plus an estimate of user time bias. Intersecting spheres in this case result in a three-dimensional volume of uncertainty. Because of the average position of the satellites with respect to the user, there is usually more confidence in a position solution in the horizontal plane (HDOP) than there is for the calculated altitude (VDOP). Additional redundant satellites provide a least-squares solution with

114   INTEGRATED NAVIGATION AND GUIDANCE SYSTEMS

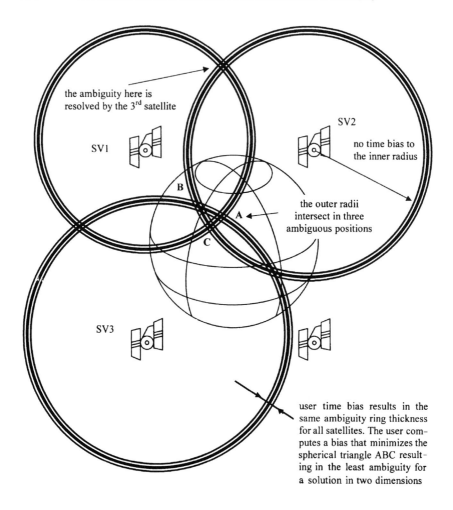

Fig. 6.2   GPS positioning from TOA in two dimensions.

added accuracy in the computed solution as well as an estimate of the covariance (uncertainty) of the position and time estimates.

A standard GPS takes a few seconds to process all of this information. This hinders it from usefulness in a real time flight application where time delay can have a very adverse impact in the cockpit. The accuracy of the resulting solution depends on user equipment error (UERE) and on satellite relative geometry with the user (GDOP). When all is said and done, you need a GDOP near or below 3 to have strong confidence in the solution.

6.1.3.2   User velocity from time of arrival.   Velocity is obtained by taking the difference between successive position measurements. This completes the navigation output from a typical GPS: position-velocity-time (PVT). Attitude is not

GLOBAL POSITIONING SYSTEM   115

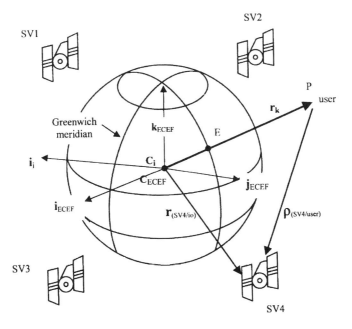

Fig. 6.3   GPS ECEF reference frame.

available from conventional processing techniques, but the information to compute it exists in the GPS carrier signal using multiple antennas. Precise velocity is not available, but Doppler velocity also can be found from the carrier frequency shift.

6.1.4   Global Positioning System Coordinate Frames

Position and velocity are computed by the GPS in Earth-centered, Earth-fixed C_{ECEF} coordinates. This coordinate frame was introduced in Sec. 2.1.1 and is shown in Fig. 6.3 where it can be seen that the user and the satellite (commonly called a SV) share the same C_{ECEF} axes, which are rotating with respect to the inertial C_i frame. The calculated position is thus expressed in C_{ECEF} coordinates, which are usually converted within the user's computer to latitude, longitude, and height above the WGS84 reference ellipsoid (see Sec. 2.2.2.1).

A code epoch pulse is transmitted every 1.5 s, and this is the typical limiting factor for how fast GPS navigation outputs can be updated. Contrast this method with an INS, which may experience a measurement update every few seconds but which has navigation outputs available to aircraft systems at rates usually above 25 Hz.

6.1.5   Fundamental Navigation Equation Set for Global Positioning System

Based on the preceding discussion, the following nonlinear vector equation can be formed to express the pseudorange measurement to each SV_i that the user has locked on to:

$$\bar{\rho} = |\mathbf{r}_{SV} - \mathbf{r}_{USER}| + cb + \bar{\varepsilon} \qquad (6.1)$$

where the equation is normally expressed in C_{ECEF} coordinates; $\mathbf{r}_{SV}$ is the SV transmitted position at the transmitted GPS time; $\mathbf{r}_{USER}$ is the user's position to be solved for at the transmitted time; c is the speed of light; b is the time bias of the user plus satellite clock offset from GPS time (this offset is known and transmitted by the satellite) plus receiver clock drift plus ionospheric delay and other system delays (from antennas and wires); and the statistical error $\bar{\varepsilon}$, or residual, of the measurement.

The smaller the statistical error $\bar{\varepsilon}$, or residual, can be made, the more accurately the user will know his position. Selective availability, discussed later, is a term intentionally added to SPS to corrupt the civilian user's solution accuracy. Ionospheric delay, because it varies with the square of the transmitted frequency, can be estimated and removed to further increase positioning accuracy if both carriers are tracked. Normally only military receivers have this capability, but multichannel GPS receivers are common in SPS systems.

The navigation equation set (NES) for GPS, requires at least four equations (from four SVs) for a full solution. More SV equations will provide a more accurate least-squares solution. The user's computer, however, does not normally solve the NES equations in the nonlinear form of Eq. (6.1) because it would take too much iteration time. The user *linearizes* the NES equation about the user's estimated current position at the time of the navigation message. This is the same procedure that produced the error correction equations for the INS, except for a very important point: the user may not have an accurate estimate of his position at GPS start-up time.

The linearized navigation equation set (LNES) is valid for the user's GPS position and time of the measurement epoch. Because it is an *error* correction equation, analogous to the INS *error* propagation equation, to initiate the solution process the user must provide the *difference between the actual pseudorange as received and the predicted pseudorange supplied by the user's computer* $\Delta \rho$. The least-squares algorithm then provides a new estimate that is re-entered into the LNES and iterated until the difference between iterated solutions is below a selected small value. At convergence the solutions, representing error corrections to estimated position and time (bias), are added to the last estimates, then converted into user coordinates (latitude, longitude, and height above WGS84), and then sent to the computer display.

It is amazing that a small device can perform all of the preceding as rapidly as it does, but consider the practical impact of linearizing about the user's estimated position at startup. If the user is not within a few miles of the estimated position at startup, the GPS receiver will not experience rapid convergence. How long it will take depends on too many variables to be discussed here, but the more incorrect the user's estimate the longer the convergence time (and possible nonconvergence). If a user is experiencing problems with the GPS at startup, the problem may not be with poor linkage with the satellites but poor initial estimates of position and time bias within his own receiver. It should be obvious, then, that an inertial navigation system navigation output may be extremely helpful to the GPS at startup time, especially in an aircraft (the estimated position, of course, may come from any avionics system with the proper interface to GPS).

6.1.5.1 Pseudorange measurements ρ.

The GPS antenna is designed to receive right-hand, circular-polarized signals. This allows antenna orientation to be relatively variable for good reception of the −160 dB watts signal. The two fundamental GPS electronic measurements are code-loop phase and carrier-loop phase.

Code-loop phase is called *pseudorange*, the most common measurement, and the difference between successive code-loop measurements is called delta pseudorange. This phase is directly related to the "time" difference between the instant in GPS time when the C/A code sequence starts (called the code epoch), $t_{\text{(sent–true gps time)}}$, and the instant when the start of the C/A code sequence is received, $t_{\text{(recvd–true gps time)}}$. It is converted to *true pseudorange* $\rho_{\text{sv/user(true)}}$ between the satellite (sv) and user's antenna by

$$\rho_{\text{sv/user(true)}} = |r_{\text{sv/user(true)}}| = c\left[t_{\text{(sent–true gps time)}} - t_{\text{(recvd–true gps time)}}\right] = c\Delta t_{\text{true}} \quad (6.2)$$

where c is light speed.

The quantity that is actually desired is the user Earth-referenced position $r_k = r_{\text{user/io(true)}}$ at the start of the epoch, where io refers to the origin of either the C_{ECEF} or C_i reference frame (recall that io = ecefo). This is the true magnitude of the user GPS antenna position vector relative to Earth's center that has been assigned the symbol r_k. In the first chapter of this work, the user Earth-surfaced referenced position was denoted ρ_K, and if the Earth-surface reference point is Greenwich, England, then

$$\rho_k = r_{\text{antenna/io}} - r_{\text{Greenwich/io}} = r_k - r_{\text{eo/io}} \quad (6.3)$$

Thus the true pseudorange $\rho_{\text{sv/user(true)}}$ from the satellite to the antenna at the start of the C/A code epoch is related to the user Earth-centered referenced position r_K by

$$\rho_{\text{sv/user(true)}} = |r_{\text{sv/io(true)}} - r_{\text{user/io(true)}}| = |r_{\text{sv}} - r_k| \quad (6.4)$$

where the space vehicle position vector at the start of the epoch r_{sv} is obtained from the data in the navigation message. The true pseudorange $\rho_{\text{sv/user(true)}}$ is related to the *measured pseudorange* ρ_{measured} (called simply the pseudorange ρ) according to

$$\rho = \rho_{\text{measured}} = \rho_{\text{sv/user(true)}} + c\left[\Delta t_{\text{atm}} + \Delta t_{\text{(user–offset)}} + \Delta t_{\text{other}}\right] + \delta_{\text{other}} \quad (6.5)$$

where the last four terms in the preceding equation are caused by atmospheric, time bias, positioning, and other errors as cataloged in later sections. Next it will be shown how the measurement for pseudorange ρ is obtained directly from the GPS receiver. It can then be used to determine the navigation solution for user position r_k with respect to Earth's center at the start of the epoch.

6.1.5.2 Pseudorange errors and time bias.

The time difference $\Delta t_{\text{(true)}}$, discussed in the previous section, times the speed of light results in true pseudorange $\rho_{\text{sv/user(true)}}$. In the absence of all error sources, $\Delta t_{\text{(true)}}$ could be obtained directly from the receiver by time-shifting the receiver's own internally generated C/A code until it correlates with the satellite transmitted C/A code sequence. When this correlation occurs, immediate knowledge is available as to which satellite is locked onto (because the C/A "gold" code for each space vehicle is relatively unique and does not correlate with the other satellite codes). Moreover, the measured time shift provides an estimate of the range from the user to that satellite (in the absence of all other errors). After the actual satellite position is obtained from the transmitted ephemeris data in the navigation message, the user Earth-referenced position r_k may then be calculated.

There is a major offset, however, in the user's clock that must be accounted for in the navigation solution. This can be expressed mathematically as

$$t_{(\text{recvd–measured gps time})} = t_{(\text{recvd–true gps time})} + \Delta t_{(\text{user–offset})} \qquad (6.6)$$

and this user offset in GPS time $\Delta t_{(\text{user–offset})}$ must be part of the navigation solution to avoid excessive error. It is often denoted by the symbol b for user time bias. It is significant that the user time bias is the same for each and every satellite that the user locks onto, for it is an inherent characteristic of a given GPS user receiver clock.

There are, of course, other significant errors that should be accounted for to maximize solution accuracy. Because the signal must travel through the atmosphere, there is a time delay Δt_{atm}. If this were the only error, a distance error δ_{atm} would result where

$$\left| r_{\text{sv/user(true)}} \right| + \delta_{\text{atm}} \stackrel{\text{atm error only}}{=} c[\Delta t_{\text{true}} + \Delta t_{\text{atm}}] \qquad (6.7)$$

When all of the other errors are included on the right-hand side of the preceding equation, the left side becomes measured pseudorange ρ according to

$$\rho = \rho_{\text{sv/user(true)}} + c\left[\Delta t_{\text{atm}} + \Delta t_{(\text{user–offset})} + \Delta t_{\text{other}}\right] + \delta_{\text{other}} \qquad (6.8)$$

which can be expressed as

$$\rho = \rho_{\text{sv/user(true)}} + cb + c[\Delta t_{\text{atm}} + \Delta t_{\text{other}}] + \delta_{\text{other}} \qquad (6.9)$$

Note that it is absolutely essential to correct for the user time bias b in the preceding equation. This is why four satellites are required for a GPS solution in three dimensions—the four unknowns are the three components of r_k and the unknown user time bias b.

The accuracy of the solution for the components of r_k and b depends on the precision of the code bits and on the extent to which the measured pseudorange ρ can be corrected by estimates for the other errors on the right-hand side of the preceding equation. For example, receiving two frequencies L_1 and L_2 allows an approximation to be made for Δt_{atm} for military users and commercial operators who may have access to both frequencies. It will be shown in later chapters that a concept called differential GPS will allow nearly all of the errors to be corrected in the preceding equation except those due to GPS receiver noise and multipath (signal reflections erroneously picked up by the receiver).

Fundamental GPS accuracy, then, depends on the inherent precision of ρ, the pseudorange measurement (the P code allows an order of magnitude measurement improvement over the C/A civilian code), and the extent that the other errors in the measurement can be minimized or eliminated.

GLOBAL POSITIONING SYSTEM

6.1.5.3 Basics of the iterated global positioning system navigation solution.
The GPS navigation solution is obtained in C_{ECEF} coordinates. The user position with respect to Earth's center in C_{ECEF} coordinates is $r_{k,\text{ecef}}$. The solution is obtained by first assuming an initial, predicted solution $(\hat{r}_{k,\text{ecef}})_0$ and then computing the vector correction $\Delta r_{k,\text{ecef}}|_0$ using a Taylor's expansion about the *assumed* initial solution. An updated position estimate $(\hat{r}_{k,\text{ecef}})_1$ is then obtained and the vector correction $r_{k,\text{ecef}}|_1$ calculated about this new estimate. The process continues until the solution converges. The basic equations are

$$r_{k,\text{ecef}} = r_{\text{user}_x} i_{\text{ecef}} + r_{\text{user}_y} j_{\text{ecef}} + r_{\text{user}_z} k_{\text{ecef}} \qquad (6.10)$$

$$(\hat{r}_{k,\text{ecef}})_{n+1} \cong (\hat{r}_{k,\text{ecef}})_n + (\Delta r_{k,\text{ecef}}|_n) + \text{HOT} \qquad (6.11)$$

where $|_n$ means the term to its left is evaluated at the nth iterated predicted value $(\hat{r}_{k,\text{ecef}})_n$ and where HOT refers to mathematical higher order terms that may be neglected in the Taylor's expansion assuming the vector correction is small.

The term $(\hat{r}_{k,\text{ecef}})_{n+1}$ in the preceding equation is the $n+1$th iterated predicted value, which, at convergence, determines the value that is actually displayed as the GPS solution. The solution for the bias b proceeds along the same lines and is given by

$$(\hat{b})_{n+1} = (\hat{b})_n + (\Delta b)_n \qquad (6.12)$$

where $(\hat{b})_0$ would be the initial estimate for pseudorange error caused by user time bias and where $(\Delta b)_n$ would be the nth iteration of the final correction to the solution for user time bias. The three components of $(\Delta r_{k,\text{ecef}}|_n)$ and $(\Delta b)_n$ can be packaged in an error vector $(\Delta x)_n$ at the nth iteration where

$$(\Delta x)_n = \left[(\Delta r_{\text{user}_x})_n \quad (\Delta r_{\text{user}_y})_n \quad (\Delta r_{\text{user}_z})_n \quad c(\Delta b)_n \right]^T \qquad (6.13)$$

Remember that the final solution $(\hat{r}_{k,\text{ecef}})_{n+1}$ must be converted to predicted latitude, longitude, and height $(\hat{\rho}_k)$ of a point on the WGS84 ellipsoid relative to Greenwich, England, before being displayed as the GPS navigation solution. The matrix setup that results in a solution for $(\Delta x)_n$ is discussed next. The following section may be skipped without loss of context in this chapter.

6.1.5.4 Linearized global positioning system navigation solution.
The matrix solution for $(\Delta x)_n$ is found using a rewritten version of the pseudorange equation for each satellite (SV). For example, if SV1 is the satellite that is locked onto, then its position would be given by its three components

$$r_{\text{sv1,ecef}} = r_{\text{sv1}_x} i_{\text{ecef}} + r_{\text{sv1}_y} j_{\text{ecef}} + r_{\text{sv1}_z} k_{\text{ecef}} \qquad (6.14)$$

and the pseudorange measurement equation for SV1 would be

$$\rho_{\text{sv1}} = \left[(r_{\text{sv1}_x} - r_{\text{user}_x})^2 + (r_{\text{sv1}_y} - r_{\text{user}_y})^2 + (r_{\text{sv1}_z} - r_{\text{user}_z})^2 \right]^{\frac{1}{2}} + cb$$
$$+ c[\Delta t_{\text{atm}} + \Delta t_{\text{other}}]_{\text{sv1}} + \delta_{\text{other}_{\text{sv1}}} \qquad (6.15)$$

A Taylor's expansion may now be applied to the scalar SV1 pseudorange measurement for the $n+1$ iteration, which results in

$$(\hat{\rho}_{\text{sv1,ecef}})_{n+1} \cong (\hat{\rho}_{\text{sv1,ecef}})_n + (\Delta \rho_{\text{sv1,ecef}}|_n) + \text{HOT} \qquad (6.16)$$

where from the application of calculus (and neglecting the HOT terms)

$$(\Delta\rho_{sv1,ecef}|_n) = \left(\frac{\partial\rho_{sv1,ecef}}{\partial r_{user_x}}\bigg|_n\right)(\Delta r_{user_x})_n + \left(\frac{\partial\rho_{sv1,ecef}}{\partial r_{user_y}}\bigg|_n\right)(\Delta r_{user_y})_n$$

$$+ \left(\frac{\partial\rho_{sv1,ecef}}{\partial r_{user_z}}\bigg|_n\right)(\Delta r_{user_z})_n + \left(\frac{\partial\rho_{sy1}}{\partial b}\bigg|_n\right)(\Delta b)_n \qquad (6.17)$$

Each of the partial derivatives is given by

$$\left(\frac{\partial\rho_{sv1,ecef}}{\partial r_{user_x}}\bigg|_n\right) = \frac{1}{2}\{[r_{sv1_x} - (\hat{r}_{user_x})_n]^2 + [r_{sv1_y} - (\hat{r}_{user_y})_n]^2$$

$$+ [r_{sv1_z} - (\hat{r}_{user_z})_n]^2\}^{-\frac{1}{2}} 2[r_{sv1_x} - (\hat{r}_{user_x})_n](-1)|_n(\Delta r_{user_x})_n \qquad (6.18)$$

$$\left(\frac{\partial\rho_{sv1,ecef}}{\partial r_{user_y}}\bigg|_n\right) = \frac{1}{2}\{[r_{sv1_x} - (\hat{r}_{user_x})_n]^2 + [r_{sv1_y} - (\hat{r}_{user_y})_n]^2$$

$$+ [r_{sv1_z} - (\hat{r}_{user_z})_n]^2\}^{-\frac{1}{2}} 2[r_{sv1_x} - (\hat{r}_{user_y})_n](-1)|_n(\Delta r_{user_y})_n \qquad (6.19)$$

$$\left(\frac{\partial\rho_{sv1,ecef}}{\partial r_{user_z}}\bigg|_n\right) = \frac{1}{2}\{[r_{sv1_x} - (\hat{r}_{user_x})_n]^2 + [r_{sv1_y} - (\hat{r}_{user_y})_n]^2$$

$$+ [r_{sv1_z} - (\hat{r}_{user_z})_n]^2\}^{-\frac{1}{2}} 2[r_{sv1_x} - (\hat{r}_{user_z})_n](-1)|_n(\Delta r_{user_z})_n \qquad (6.20)$$

$$\left(\frac{\partial\rho_{sv1,ecef}}{\partial b}\bigg|_n\right) = c(\Delta b)_n \qquad (6.21)$$

If the predicted distance from SV1 to the user antenna for the nth iteration is designated $(\hat{\rho}_{SV1})_n$, the *predicted pseudorange to SV1*, where

$$(\hat{\rho}_{sv1})_n = \{[r_{sv1_x} - (\hat{r}_{user_x})_n]^2 + [r_{sv1_x} - (\hat{r}_{user_x})_n]^2 + [r_{sv1_z} - (\hat{r}_{user_z})_n]^2\}^{\frac{1}{2}} + c(\hat{b})_n \qquad (6.22)$$

then a vector form of the equation for $(\Delta\rho_{sv1,ecef}|_n)$ may be constructed as

$$(\Delta\rho_{sv1,ecef}|_n)$$

$$= \left[\frac{-[r_{sv1_x} - (\hat{r}_{user_x})_n]}{(\hat{\rho}_{sv1})_n} \quad \frac{-[r_{sv1_x} - (\hat{r}_{user_x})_n]}{(\hat{\rho}_{sv1})_n} \quad \frac{-[r_{sv1_z} - (\hat{r}_{user_z})_n]}{(\hat{\rho}_{sv1})_n} \quad 1\right](\Delta x)_n$$

$$= [(\hat{\beta}_{11})_n \quad (\hat{\beta}_{12})_n \quad (\hat{\beta}_{13})_n \quad 1](\Delta x)_n \qquad (6.23)$$

where the three beta elements in the row vector are the negatives of the predicted values for the direction cosines of $(\hat{r}_{sv1/user})_n$ with respect to the user for the nth iterated solution and where $(\Delta x)_n$ has been defined by Eq. (6.13).

The left-hand side in Eq. (6.23) is approximated for each iteration by

$$(\Delta\rho_{sv1,ecef}|_n) \cong \rho_{sv1,ecef} - (\hat{\rho}_{sv1,ecef})_n \qquad (6.24)$$

where $\rho_{sv1,ecef}$ is the actual GPS pseudorange measurement for SV1 for the epoch. This is then repeated for at least four of the satellites, resulting in a 4-by-1 measurement error vector $\Delta\rho_{sv,ecef}|_n$ and the following vector-matrix equation:

$$(\Delta\rho_{sv,ecef}|_n) = \begin{bmatrix} (\hat{\beta}_{11})_n & (\hat{\beta}_{12})_n & (\hat{\beta}_{13})_n & 1 \\ (\hat{\beta}_{21})_n & (\hat{\beta}_{22})_n & (\hat{\beta}_{23})_n & 1 \\ (\hat{\beta}_{31})_n & (\hat{\beta}_{32})_n & (\hat{\beta}_{33})_n & 1 \\ (\hat{\beta}_{41})_n & (\hat{\beta}_{42})_n & (\hat{\beta}_{43})_n & 1 \end{bmatrix} (\Delta x)_n = (\hat{H})_n (\Delta x)_n \quad (6.25)$$

The preceding equation is solved using the least-squares approach for

$$(\Delta x)_n = \left\{ (\hat{H})_n^T (\hat{H})_n \right\}^{-1} (\hat{H})_n^T (\Delta\rho_{sv,ecef}|_n) \quad (6.26)$$

which is added to the previous predicted values for $(\hat{r}_{k,ecef})_n$ to obtain $(\hat{r}_{k,ecef})_{n+1}$. These, in turn, allow an updated prediction to be made for the pseudorange $(\hat{\rho}_{sv1,ecef})_{n+1}$ and the resulting calculation of the new measurement error $\Delta\rho_{sv1,ecef}|_{n+1}$. Then, with the recomputed measurement matrix $(\hat{H})_{n+1}$ a solution can be found for $(\Delta x)_{n+1}$, et cetera.

The principles of linear algebra do not limit the rows of the predicted measurement matrix to four, but only require that the rows be linearly independent. Notice how the direction cosines of the line-of-sight vectors from user to each space vehicle in turn determine each of the row vectors in the predicted measurement matrix. The relative linear independence of each row is therefore determined by satellite geometry with respect to the user position. For a given number of satellites in view (which determines the number of rows), this relative linear independence determines the quality of the resulting solution. The formal name given to this quality metric is the geometric dilution of precision (GDOP).

6.1.5.5 Geometric dilution of precision from the linearized navigation solution. Comparing Eq. (6.26) with Eq. (4.44) in Sec. 4.3, we see that this is a standard nonweighted least-squares problem with solution

$$\left(H_k^T H_k \right)^{-1} H_k^T \Delta z_k = \Delta\hat{x}_k(+) \quad (6.27)$$

which can be compared with the weighted least-squares solution given by

$$\left(H_k^T R^{-1} H_k \right)^{-1} H_k^T R^{-1} \hat{z}_k = \hat{x}_k(+) \quad (4.44)$$

It will be left to the reader to ponder about the possibility of a sequential solution to the LNES for GPS following the reasoning pattern that resulted in the Kalman filter iterative solution to the weighted least-squares problem.

Recall that the solution to the least-squares problem provides an estimate of the covariance for the residuals, and it has been already been noted that GPS GDOP is strongly influenced by geometry, which in the preceding equation set is represented by the rows of direction cosines.

Because the rows of the *H* matrix represent estimated line of sights from user to satellites, it is apparent that if two or more satellites line up with the user that the rows become linearly dependent and the solution will break down. This will

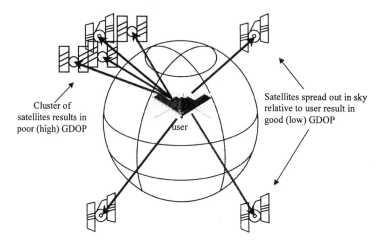

Fig. 6.4 Satellite geometry and GDOP.

cause a large increase in the covariance and resulting GDOP. GDOP now can be defined

$$\text{GDOP} = \sqrt{\text{tr}(\boldsymbol{H}^T \boldsymbol{H})^{-1}} \qquad (6.28)$$

where tr indicates the "trace" or sum of the diagonals of the geometry solution matrix $(\boldsymbol{H}^T \boldsymbol{H})^{-1}$. From the above definition $(\text{PDOP})^2 = (\text{HDOP})^2 + (\text{VDOP})^2$ and $(\text{GDOP})^2 = (\text{PDOP})^2 + (\text{TDOP})^2$.

Thus, *satellite geometry relative to the user has a very powerful impact on the accuracy of the solution.* This is why the vertical dilution of precision VDOP is usually worse than the horizontal HDOP: satellite pseudorange measurements are relatively insensitive to changes in user height above the WGS84 ellipsoid unless satellites are directly overhead. Conversely, if too many satellites are overhead, then the HDOP will degrade because the insensitivity now will be relative to user motion in the local level plane. This is illustrated in Fig. 6.4.

6.2 Global Positioning System Segments

6.2.1 *Global Positioning System Space Segment*

6.2.1.1 Global positioning system satellite blocks. There have been three phases of satellite development through 1997, and each phase has a block of satellites. Two more are planned following 1997. The first phase developed initial concept validation satellites (Block 1), and they no longer fly. The second phase produced operational satellites (Block II) and upgraded operational satellites (Block IIA) of which 24 were in service in 1997. The two other phases are for replacement satellites (Lockheed-Martin Block IIR through 2004 and Rockwell International Block IIF afterwards). SVs are designed for a 10-year life and must meet by contract a 7.5-year operational period.

Block IIR satellites. The first attempt to launch a IIR satellite resulted in an explosion of the Delta rocket on January 17, 1997. On July 22 of that year, a

GLOBAL POSITIONING SYSTEM

Delta 2 booster ($55 million) successfully launched a Lockheed Martin IIR GPS satellite ($40 million) from Cape Canaveral, Florida. The IIRs are computer reprogrammable from the ground, have enhanced redundancy management, and provide frequency-hopping cross links to communicate with other IIR satellites. This will result in an autonomous navigation capability (AutoNav) for 180 days in the unlikely event that communication is lost from the master ground station. In addition, the spacecraft processing unit that manages satellite functions is fully capable of being reprogrammed from the ground.

The IIR satellites and their associated ground control software are designed to be immune to the year 2000 millenium crossover problem, an event that is feared to create a "divide by zero" frenzy in some of the world's computers. GPS satellites are more likely to be affected by an end-of-week (EOW) rollover affecting GPS clock systems. This is discussed further in Sec. 6.2.1.2 on the navigation message.

By the year 2006 the IIR satellites will remove selective availability, the intentional degradation of the transmitted signal to limit user accuracy, and handheld units may then see accuracy within 30 ft CEP. The military will retain an option to deny GPS coverage to specific parts of the world, to prevent enemies from using it to target missiles. The new satellites and receivers will also allow all users to receive simultaneous signals to compensate for ionospheric distortion (see Sec. 6.2.1.2—Additional civil frequencies).

Block IIF satellites. It will take up to 20 years to replace a generation of satellites, and the Block IIF SVs will begin replacing the Block IIR by 2004. The planned lifetime of these satellites is 15 years, and a constellation of up to 30 SVs from the planned 24 may result. They will have a *spot beam* capability for geographic region and operate at higher power levels to resist jamming. The goal of these satellite designs is to improve civilian support and accuracy while denying GPS to hostile forces, wherever they may occur.

A new generation of satellites is under consideration called GPS III, satellites intended for the year 2015 and beyond. This system will reach for positioning goals in the millimeter range for positioning and the picosecond level for timing. This will require code tracking in the receivers as discussed in Chapter 8.

6.2.1.2 Transmitted signals. The purpose of the navigation payload in these satellites is to transmit the ranging codes and navigation message to the user community [called the *total navigation payload* (TNP)]. The code generator generates the C/A code and the military precision P(Y) codes for the navigation message that is transmitted by an L-band subsystem on two frequencies (simultaneously by all satellites) called L_1 (1575.42 MHz) and L_2 (1227.6 MHz). The two frequencies allow the time delay caused by the ionosphere to be detectable by a capable receiver. The Y in P(Y) indicates that the military codes are encrypted and that users require a key to receive the P code. This prevents "spoofing" by hackers (and others).

The code for standard positioning service is a C/A code (a Gold code of register size 10) that has 1023 bits in its sequence at a clock rate of 1.023 MHz and a code period of 1 ms. The code contains the navigation message in a 50 bit/s data stream of 30 bit words and "randomly" phase modulates the L_1 carrier signal, "spreading" the spectrum over a 1 MHz bandwidth (see Fig. 6.5). The navigation message contains ephemeris and satellite clock correction data and hand-over information

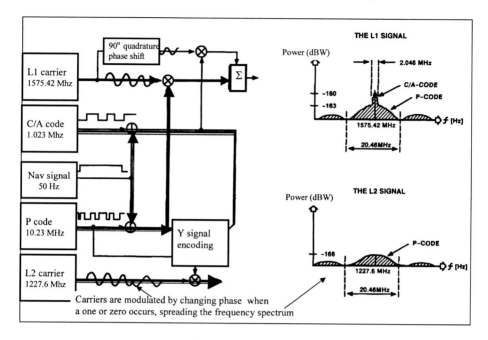

Fig. 6.5 GPS signal characteristics.

that military users can use to acquire the precision (P) code and thus obtain better performance (specialized receivers exist that directly track the P code). Recall that there is a different C/A code (member of the "Gold" code family) PRN for each SV, and GPS satellites are identified by their PRN number.

Standard positioning service and precise positioning service. The C/A code enables a user to obtain standard positioning service (SPS) with an advertised accuracy of 100 m (two sigma). All users must receive this code to navigate with GPS. The P code enables authorized users to obtain precise positioning service (PPS), which is approximately an order of magnitude better in positioning accuracy.

Frequencies L1 (1575.42 MHz) and L2 (1227.6 MHz) are multiples of the precision code "chipping rate" of 10.23 MHz. L1 and L2 are 154 and 120 times, respectively, the chipping rate and were designed that way so that the codes could be efficiently reproduced within every GPS receiver. The L1 frequency contains both the C/A and P codes—recall that if the P code is encrypted by the military it is called the Y code ("anti-spoofing" mode). The L2 frequency, however, only contains the P (or Y) code. The encryption has two keys, one with an annual access to authorized users and one with a weekly access that allows special operations to bypass locking onto the C/A code.

SPS users can easily lock onto the C/A code, with a repeat period of 1 ms. For PPS users the task of locking onto the P code (10.23 MHz rate) is much more difficult, and so the navigation message contains a "hand-off word" (HOW) to assist subsequently lock of the P (or Y) code. Note that the navigation message is available whenever the C/A code is being tracked by the GPS receiver.

Navigation message. The 50-Hz navigation message requires 12.5 min to be fully received by the user. However, parts of the message that are vital to GPS navigation are repeated in each "frame" of data. Each frame of data is 1500 bits long divided up into five 300-bit subframes (a subframe is thus obtained every 6 s). The first three subframes of each frame always include an 8-bit preamble (10001011) and "handover" or "handoff" word, followed by satellite position and clock correction data. The remainder of each frame is atmospheric correction data and almanac data that does not repeat until the 12.5-min navigation message is complete.

The GPS clock systems count 1023 weeks in a binary format from the "GPS start time" of 0000 GMT January 6, 1980. The handoff word in the navigation message contains the current week. Time is further broken down into 6-s intervals by the message. The rollover to zero weeks at 0000 GMT August 22, 1999, occurs at a 19-year interval. The navigation message is full of time correction parameters that help the user receiver calculate the precise corrections necessary for a useable PVT solution.

The clock error parameters from the navigation message are some of the inputs to the data processor that solves for PVT using PR measurement and dead-reckoning measurements (if available). The solution to the four-equations in four-unknowns positioning problem is typically solved with a Kalman filter implementing the weighted least-squares solution. The dynamics model is a set of dead-reckoning equations plus clock bias and drift terms. It is used to provide estimates of pseudorange at update time and weighting any other measurements that may be available to update its state estimate.

The Kalman filter gains minimize the residual covariance between the predicted and actual pseudorange and update the PVT estimates from the dead-reckoning model. The model then propagates the state to the next update time, operating in much the same fashion as the Kalman filter equations for the INS.

Additional civil frequencies. Because of the rapidly expanding use of GPS, there is concern over the encoding of the L2 signal by the military, thus denying its use for civilian applications. In March of 1998, Vice President Al Gore announced that the Interagency GPS Executive Board selected the GPS L2 frequency as the spectrum location for a second civilian signal called L5. In addition he promised a third civil signal with structure and frequency to be specified in time for placement on the seventh Block IIF satellite due for launch by Boeing in year 2004.

The government will identify one of the new signals as a *safety-of-life service signal* for use in air traffic control. Such a frequency must be clear of interference, and thus the decision has both national and international implications. For example, other programs competing for spectrum allocation include the U.S. Joint Tactical Information Distribution System (JTIDS), Russia's GLONASS satellite navigation system, some radar bands, and a European proposal analogous to JTIDS called the Multifunctional Information Distribution System (MIDS).

Dr. James J. Spilker of Stanford Telecom has proposed a *G prime option* that would put the new civil frequencies in the "null" of the military L2 signal, thus reusing bandwidth and promising accuracy improvements. The GPS Joint Program Office (JPO) is testing candidate signal structures for the new frequencies because a recommendation must be made in 1998 to meet the 2004 satellite launch date, and Dr. Spilker's proposal is a leading candidate with much promise.

The payoff for providing these new frequencies is difficult to estimate, but if civilian innovation using L1 and L2 is any indicator, it will be considerable. Precise real time positioning may become commonplace for guiding individual vehicles (snowplows, mining equipment, airport ground traffic), controlling fleets of commercial vehicles (land, sea, or air), and automating emergency response networks. The technical obstacles pale in comparison to the political ones, however, because of the level of cooperation required (and the resulting funding issues) between national and international governmental agencies.

6.2.1.3 Transmission hardware on the space vehicle. The total navigation payload (TNP) hardware generates the signals that provide the navigational capability of the GPS. The antennas are always pointed toward the Earth to ensure uniform reception of navigational signals. The cross-link transponder and data unit (CTDU) provides direct satellite-to-satellite communications that permit autonomous operation for 180 days (i.e., without ground control). The mission data unit (MDU) integrates all ephemeris calculations, encryption, and pseudorandom code generation. The MDU software is Ada running on a MIL-STD 1750A 16 MHz processor.

6.2.2 Global Positioning System Control Segment

The operational control segment (OCS) keeps the satellites in correct orbits (station keeping) and monitors satellite operation. It updates each satellite's clock and ephemeris in the navigation message. The ephemeris parameters are a precise fit to the GPS satellite orbits and are valid up to 6 h from last upload. The master control facility is located at Falcon Air Force Base, and *this facility is necessary to maintain long-term GPS operation.* The other control stations, located in Hawaii, Ascension Island, Diego Garcia, and Kwajalein, are data transmission and reception sites (see Fig. 6.6). The upload corrections for each satellite are transmitted daily back to the satellites from Ascension, Diego Garcia, and Kwajalein on S-band. Corrections are accomplished with a master Kalman filter.

The navigation message data can be stored for 14–210 days. Almanac data is used to predict satellite position and aid in satellite signal acquisition. The OCS resolves satellite anomalies and makes pseudorange measurements at the remote monitor stations to update GPS time. This is a complicated matter involving relativity effects, frequency drift, and orbit estimation that results in time accuracy of a few nanoseconds and orbit accuracy within a few meters across the entire GPS constellation.

6.2.3 Global Positioning System User Segment

The revolution in GPS technology is largely because of the innovations in the user community. Users are guaranteed only -160 dB W of power near the Earth's surface from the carrier that is associated with frequency L_1 (-166 dB W for L_2 precision users). This resulted in the initial sets of the 1970s being very large and costly devices procured by the military with loads of analog circuitry and only two channels. The largest unit known to the author was a 70-lb GPS used on a ship.

GLOBAL POSITIONING SYSTEM

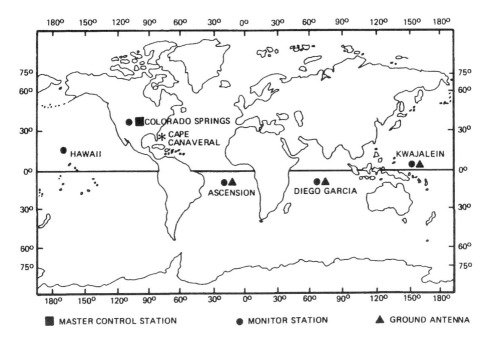

Fig. 6.6 GPS control stations.

In 1999 the 12-channel receiver is digital, weighs a few pounds, and is relatively inexpensive. GPS receivers may be as small as a wristwatch, like the one Capt. Scott O'Grady wore when he was rescued in Bosnia in 1995. The channels are the means to lock on to a satellite. A single channel GPS receiver must cycle from one satellite to the next to get each navigation message—this would be an example of *sequential tracking receivers*. Multichannel sets can simultaneously track many GPS satellites (as long as they are in view). A military five-channel set can parallel process the navigation message simultaneously from four satellites while receiving the navigation message from a fifth satellite.

The GPS user equipment measures the pseudorange, which is the GPS time of arrival plus bias, computes the position of the satellites it is receiving from the navigation message, then determines the least-squares solution to determine three components of position in C_{ECEF} coordinates. As described in the preceding sections, the solution automatically yields the dilution of precision accuracies.

Military aircraft (1999) use the Miniature Airborne GPS Receiver (MAGR), or Standard GPS IIIA receiver, which has three different modes: Doppler radar system (DRS) with Doppler radar aiding, the INS mode, and the position-velocity-acceleration (PVA) mode (no aiding). It is a continuous tracking receiver that uses the fifth channel to read the navigation message of an additional satellite. This receiver operates in high dynamic environments (but remains limited by the low output rate of its navigation solution). Multiplex receivers exist that cycle up to 50 Hz between satellites while reading data from all of the satellites in view.

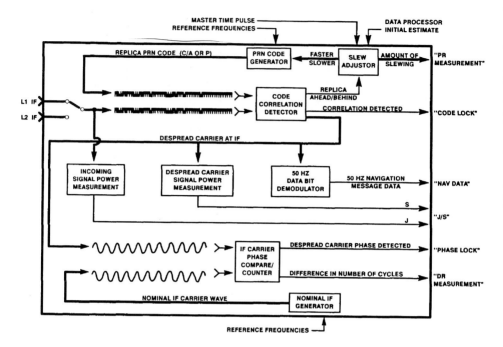

Fig. 6.7 GPS receiver characteristics.

A block diagram for a generic GPS receiver from the NAVSTAR GPS JPO is shown in Fig. 6.7. The receiver generates a copy of the C/A or P code based on a *master time pulse* from the quartz clock. The copy is then slewed by the expected pseudorange value resident in the receiver at initialization. When the copy exactly matches the desired pseudorange code in the intermediate frequency (IF), a *correlation lock-on* occurs. The jammer-to-signal (J/S) ratio in Fig. 6.7 is particular to military receivers. If in a jamming environment, the receiver can be programmed to decide whether L1 or L2 carrier will be tracked. In a commercial receiver the carrier power to the noise level would be measured.

Regardless of the type, a GPS receiver basically implements a phase-lock loop circuit to extract the pseudorange and delta-range of a selected satellite. The phase-locked loop locks on the phase of the carrier frequency and is therefore able to detect any phase change of the carrier. This property has been shown to have significant potential because carrier tracking allows navigation solutions in the centimeter range at a high solution output rate. These applications are discussed in Chapter 8.

6.2.3.1 Global positioning system flight navigation.   The FAA Advisory Circular AC 90-94 (December 14, 1994) was the first attempt to introduce standard positioning service (SPS) GPS to the aviation community. The International Civil Aviation Organization (ICAO) considers GPS part of the Global Navigation Satellite System (GNSS), which also includes the Global Orbiting Navigation Satellite System (GLONASS) developed by Russia.

Although it is the responsibility of the pilot in command to ensure that GPS is legal in a given country, the bottom line is that an appropriately certified GPS may be used under IFR rules as a long-range navigational system (LRNS) if certain conditions are met. The LRNS must ensure that either an approved technique of receiver autonomous integrity monitoring (RAIM) is employed or that an alternate means of navigation [such as an INS or long-range navigation (LORAN)] is available and monitored in the absence of RAIM.

RAIM, discussed in more detail later, is a method the GPS receiver and software uses to check on the reliability of the satellite signals that it is receiving. For example, if more than four satellites are being received, comparative solutions may be generated to flag bad satellite data and to possibly identify the offending satellite. RAIM is *not* the GDOP, a concept that depends solely on the geometry of the satellites with respect to the user. RAIM should advise the user when system integrity is bad, even when GDOP is low. GDOP will not inform a user that SV data is corrupted.

GPS "overlay approaches" are allowed where published as long as the original NAVAID is operational, and GPS-only approaches are allowed when RAIM is available. A typical GPS approach plate used at NASA Ames Flight Research Center is depicted in Fig. 6.8. This is done via waypoints that are accurate to within 0.01 arc min and permanently programmed into the GPS by an FAA-approved source. A receiver must have a minimum of nine waypoints programmed, each consisting of latitude, longitude, and a name. These waypoints are normally the initial approach fix, final approach fix, and missed approach point along with any intermediate and holding fixes. The instrument approach procedure must be retrievable from the programmed navigation database, and the user may not change database entries. The user may, however, create "USER waypoints" as desired.

Navigation output is displayed in various forms and formats to the pilot, including moving map software displays for laptop computers. There are no advisory circulars or certified display technologies that must be followed. The pilot as always is responsible for situational awareness, and a GPS is never a substitute for good sense or lack of clutter in the cockpit. A strong lesson in this area is provided by the grounding of the GPS-guided passenger ship *Royal Majesty* on June 10, 1995, near Nantucket, Massachusetts. The ship was 17 miles off course, but the display appeared totally normal due to the use of a smoothing algorithm that integrated the GPS and dead-reckoning solutions. The ship's crew believed the display.

Precision approaches using GPS have been accomplished and will be discussed in more detail in Chapter 8. For these approaches the aircraft locks onto the signal from pseudolites, or GPS *integrity beacons*. These are Earth-surface-fixed GPS transmitters that mimic a GPS satellite. They are used because GPS does not normally have either the system integrity or the system availability that is required for precision approaches. However, pseudolites can jam user receivers inadvertently when the users are too near the beacon, and because of their very low power their signal is difficult to receive at a long distance. This is called the *near-far* pseudolite problem.

The FAA is confident enough of GPS potential, however, to have ended all work on the microwave landing system (MLS), a concept designed to replace the

130   INTEGRATED NAVIGATION AND GUIDANCE SYSTEMS

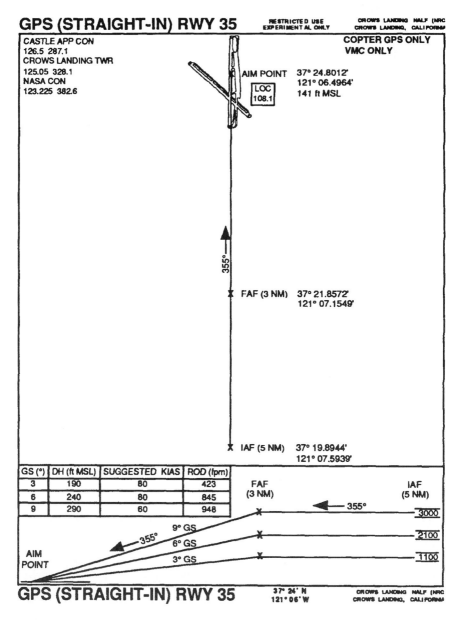

Fig. 6.8   GPS approach plate.

conventional ILS but which was preempted by GPS. However, it should be noted that outside the United States, and especially in Europe, the MLS has met more acceptance and success.

6.2.3.2 Global positioning system susceptibility to jamming. In upstate New York from December 30, 1997, through January 12, 1998, approximately a dozen GPS-equipped commercial aircraft filed interference reports due to a loss of the GPS signal, with one Continental flight on December 30 reporting a total loss of GPS. The cause was inadvertent jamming from a GPS antenna test at an Air Force laboratory in Rome, New York, in which a computer failed to terminate a low power transmitter stuck at the GPS carrier frequency 1227.6 MHz. The problem was not terminated until laboratory personnel read news reporting a loss of GPS in their geographic area.

GPS jammers are available on the open market. A Russian company has displayed 4 W GPS jammers at the Moscow Air Show that can disrupt GPS out to 100 n miles. This incident caused the FAA to declare an interference zone of 150 n miles and re-emphasized the DoD policy that approval is needed from the Joint Chiefs of Staff to perform a GPS jamming test. The President's Commission on Critical Infrastructure Protection has specifically cited GPS vulnerability to jamming. Because of this, there is little confidence that GPS will ever become a sole means of navigation in the air.

The encoded Y-code does not protect against jamming, only against *spoofing*, the intentional transmission of a false GPS signal. The basic effect of jamming, or intentional radio signal (RF) interference, is to reduce the carrier to noise ratio toward the tracking threshold where the receiver loses lock and thus its navigation capability. RF interference detection is typically done in the receiver by monitoring the automatic gain control, which will increase above the ambient thermal level if interference exists. This does not require satellite tracking lock-on.

The carrier and code tracking loops may be enhanced to resist jamming. This is normally done by narrowing the predetection bandwidth of a receiver, a process that increases the chance of break-lock due to aircraft maneuvering. The carrier aiding of the code loops that is accomplished to improve navigation accuracy does not increase resistance to jamming, and it is the weak link in a GPS with no external aiding. If the navigation solution is implemented with a Kalman filter, then the filter, when jamming noise is sensed, can weight the propagated dead-reckoning state estimate higher and de-weight the noisy measurement. The Kalman filter thus can adapt to jamming to shield the PVT solution from external noise. If the PVT solution is implemented only by a least-squares simultaneous solution of equations, it will become very erratic as noise increases.

Because of the many fears surrounding the GPS jamming problem, it is wise to have alternate means of navigation available. The U.S. Army Space and Missile Defense Command, for example, in 1998 tested a commercial, off-the-shelf car navigation system in Army tactical vehicles that would operate autonomously if jamming or spoofing of the GPS signal occurred. In April 1998, 15 computer hackers between 19 and 29 years of age broke into a Pentagon network, pirated software that coordinates the military's GPS, and threatened to sell it to terrorists. The hackers, calling themselves the "masters of downloading," were from the United States, Great Britain, and Russia.

6.2.3.3 Multisensor navigation. GPS is considered a "sensor" when integrated with other navigation aids into a flight management system (FMS). These multisensor units provide improved accuracy, integrity, and availability. GPS is normally combined with LORAN in general aviation aircraft, and with INS in military aircraft. Advanced integration schemes usually rely on the Kalman filter. The basic idea of external aiding enhancements is to enable GPS to provide an over-determined navigation solution. This increases the ability of GPS to maintain lock in the presence of RF interference. This aiding can occur from a variety of sources, such as an inertial navigation system, Doppler radar, CADC inputs, or magnetic compass sensors.

A very effective synergy exists with the GPS/INS systems in a tightly coupled configuration. GPS initializes and calibrates the INS and limits its long-term error, while the INS can remove most of the vehicle dynamics from the GPS tracking loop and permit a narrow receiver bandwidth for noise rejection. More of the benefits from the marriage of these systems is covered in the next chapter.

If a Kalman filter is used for implementing the GPS navigation solution using dead-reckoning equations, a modeling problem occurs when acceleration, an "inferred" state, is not constant over the propagation interval. This may occur, for instance, during rapid maneuvers. The rate of change of acceleration, or "jerk," imposes a serious modeling error that degrades the GPS position accuracy. The degradation is directly proportional to the magnitude of jerk and inversely proportional to the number of measurements available during the update interval.

Thus many Kalman filter designs for GPS processors allow external source aiding by an IMU, Doppler velocity, barometric altimeter, and atomic clocks. The operator may be able to set special modes, such as when the unit is stationary (survey mode) or at a constant altitude ("coasting"). These aids effectively augment the dead-reckoning equations and significantly improve the dynamics model used by the filter. The use of multisensor technology to improve the PVT solution is called *dynamic aiding*, and it improves the receiver's resistance to jamming. This is because the tracking loops can increase their bandwidth (i.e., sensitivity to change) to hold tightly onto signals without unmodeled accelerations causing loss of lock. Moreover, during periods of low dynamic activity, the GPS can use an external-aiding source (such as altitude) in lieu of a pseudorange and operate accurately using only three satellites, a process called *coasting*.

6.3 Selective Availability

The largest error in GPS is selective availability (SA), induced by the military to intentionally degrade the navigation solution. Because it is an induced error that is an extreme irritant to many commercial users, it may have a limited lifetime in today's rapidly changing environment. SA error is discussed separately from other GPS errors in this text. SA may consist of the transmission of intentional orbital errors, called ε_0, and the dithering of the satellite clock, called δ. The clock error has periods between 4 and 12 min with accompanying pseudorange variations up to 70 m. An example of the clock drift errors for SV14 is provided in Fig. 6.9, courtesy of Orion Dynamics and Control, Boulder, Colorado.

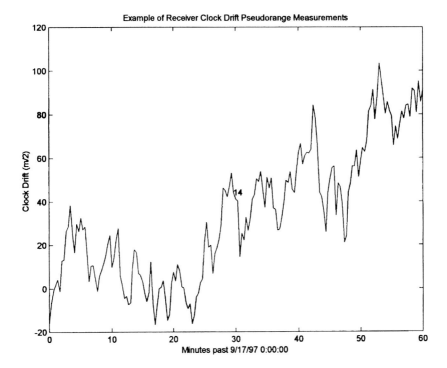

Fig. 6.9 Example of clock drift (courtesy of Orion Dynamics and Control).

SA is "spatially correlated," which means in practical terms that it can be modeled and predicted (like the stock market averages). Work at Ohio University showed that SA could be modeled within 10 m for up to 200 s before becoming seriously degraded. The dominant SA error causes random-like errors up to 30 m with approximately a 3-min period. Officially, the performance of GPS without SA should be in the 20-m range for horizontal positioning accuracy (95%).

6.3.1 Why Is Selective Availability There?

Field tests during the 1970s resulted in SPS performances that were much better than anticipated. The C/A code provided 20–30 m actual accuracy rather than the predicted accuracy of no better than 100 m. The DoD decided (1983) to intentionally degrade the accuracy to 100 m. This level of accuracy was chosen because it was comparable with that of a VOR during a nonprecision approach.

6.3.2 Impact of Selective Availability on Global Positioning System

There is considerable public pressure to release frequency L2 to the general public and to implement a new L5 frequency for military use. The code encryption would be taken off of L2 and placed on L5. This is because of the enormous amount of investment that commercial users have made in obtaining commercial L1/L2 geodetic receivers (primarily for surveying). The government would have to consider the costs of switching military equipment to handle L5 and compare it with the consequences of denying civilians the use of L2 for precision applications. Frequencies L3 and L4 are assigned classified missions on GPS satellites.

It is ironic that the intentional degradation of the navigation message to degrade performance has caused innovative responses resulting in enormous performance improvements. The innovation within the user community, fueled by frustration at the intentionally degraded accuracy of standard GPS units, has produced a new generation of navigation hardware that is not only no longer susceptible to SA, but which far exceeds the advertised GPS performance without SA. In March 1996 a Presidential Policy Directive recommended turning off SA by the year 2006.

6.4 Global Positioning System Error Sources and Error Modeling

Errors affecting the GPS come from the control, space, and user segments. Errors are normalized into equivalent pseudorange error so that error budgets use a standard evaluation measure of impact. This results in the user-equivalent range error (UERE), which is the RSS of all of the error sources. The net expected error is the UERE times the GDOP, which we have seen depends primarily on geometry. The error characterization using GDOP was discussed in Sec. 6.1.3. Recall that GDOP was best (approximately 2) when the satellite geometries were widely dispersed about the user's position.

It has been generally shown that, except for SA, the ionosphere is the dominant source for GPS error (4 m SPS, 1 m SA). This is because the free electrons in the ionosphere, the first atmospheric layer encountered by a transmitted signal, refract the signal and cause a delay that depends on signal frequency. This refraction varies with time and position and is especially significant following solar geomagnetic storms and solar flares. Dual-frequency receivers have the capability to significantly reduce ionospheric delay by computing the error correction in real time.

The next two atmospheric layers, the stratosphere and the troposphere, cause a nondispersive delay that is not signal frequency dependent. This delay has a dry atmospheric part that can be modeled if barometric pressure is known, and a part due to humidity that is highly variable and especially sensitive to the passage of weather fronts. The net result is a degradation of GPS computed height that is typically a factor of two worse than the GPS calculations for latitude or longitude. This degradation in HDOP is called *tropospheric delay*, but in actuality both the stratosphere and the troposphere cause the delay.

The error budget for other sources is summarized in Table 6.2. The fundamental rms error of the GPS solution, driven by both geometry and user errors, is given by

$$\sigma_{\text{RMS}} = \sigma_{\text{UERE}}(\text{GDOP}) \tag{6.29}$$

where σ_{UERE} is the root sum squared (RSS) of the user error sources as shown for each column of Table 6.2, and GDOP is the square root of the trace of the geometry solution matrix $(H^T H)^{-1}$ as given by Eq. (6.28) when the solution converges. Thus σ_{RMS} is the value for overall accuracy that can be expected from GPS given satellite geometry, user equipment, and environmental factors.

Table 6.2 Global positioning system error budget

Source	SPS, one σ: meters	SA off, one σ: meters
Ephemeris	2.1	2.1
Clock	20	2.1
Ionosphere	4	4
Troposphere	0.7	0.7
Multipath	1.4	1.4
Receiver	0.5	0.5
σ_{UERE}	20.6	5.3

6.5 International Satellite Navigation Systems

The GNSS is a concept sponsored by the international telecommunication industry called Inmarsat to provide geostationary satellites that will augment GPS and GLONASS. These satellites will provide a service called Egnos (GNSS-1) that will link GPS and GLONASS satellites with transponders and provide a worldwide nonprecision approach capability with a focus on Europe. A navigation service is planned with a focus on Asia by Japan (GNSS-2).

GPS and GLONASS satellites are not synchronized (GLONASS time includes leap seconds), and a primary difference between GPS, GLONASS, and the European satellites is in the radio transmission technology. GLONASS uses frequency division multiplex, GPS uses code division multiplex, and the European satellites plan to use time division multiplex. A user planning to lock onto whichever satellites are available thus has a considerable challenge in designing a receiver. On the other hand, as can be seen from Fig. 6.10, more satellites are available when both systems are used for navigation.

6.5.1 GLONASS

The Russian GLONASS system evolved from a Doppler measurement satellite system called Cicada in a manner similar to the evolution of GPS from the Navy Transit satellite system. GLONASS is planned to have 24 satellites, eight per orbital plane, with the planes separated by 120 deg. As with GPS, four satellites are required for a navigation solution based on time of arrival. GLONASS does not intentionally degrade the transmitted signal as the United States does with SA.

The first satellite (named Uragan or "hurricane") was launched in 1982, and the tenth Block I prototype was launched in 1985. Forty-three Block IIv followed through 1996. As of January 1998, GLONASS has 16 active satellites, but no launches have occurred since a December 1995 launch of a Block IIc satellite with a design life of three years. A master control center and tracking stations in Russia control data uploads to the satellites twice per day.

Position estimates from GLONASS (10 m UERE rms) are better than GPS in the presence of selective availability (25 m UERE rms), but GLONASS uses the SGS85 (PZ-90) Earth ellipsoid for navigation, not the WGS84 used by the United States. The conversion between the two systems in use by the Lincoln Laboratory

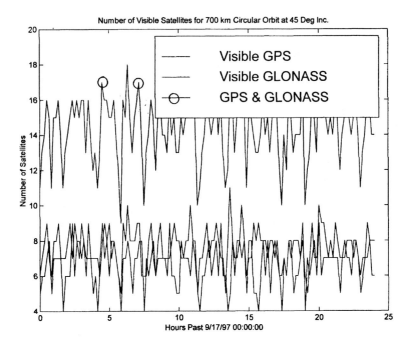

Fig. 6.10 GPS visibility with GLONASS (courtesy of Orion Dynamics and Control).

at the Massachusetts Institute of Technology is (in meters)

$$\begin{bmatrix} x \\ y \\ z \end{bmatrix}_{\text{WGS-84}} = \begin{bmatrix} 0 \\ 0 \\ 4 \end{bmatrix} + \begin{bmatrix} 1 & 3(10)^{-6} & 0 \\ 3(10)^{-6} & 1 & 0 \\ 0 & 0 & 1 \end{bmatrix} \begin{bmatrix} u \\ v \\ w \end{bmatrix}_{\text{SGS 85}} \quad (6.30)$$

GLONASS does not use pseudoranging codes but does employ spread-spectrum radio channels that transmit an epoch of data every millisecond in the L-band of frequency. The transmissions are frequency multiplexed between 1240–1260 MHz and 1597–1617 MHz with approximate 0.5 MHz channel spacings that allow 25 channels. The GLONASS navigation message is 511 bits long at a 50 baud rate that, like the GPS navigation message, is modulated using binary phase-shift keying. The Russians have been under pressure by radio astronomers and the International Telecommunications Union (ITU "frequency police") to reduce the frequencies to 12 primary channels plus two spares. This it can do by using the same frequencies for those satellites separated by 180° of latitude.

6.5.2 European Geostationary Navigation Overlay Service

The European Geostationary Navigation Overlay Service (EGNOS) is a set of navigation payloads on board geostationary satellites that are monitored by Europeans. It is intended to provide services to civil users under civil operation and control by the year 2010. Its goal is establishing a satellite positioning service for Europe as part of the development of GNSS. GNSS-1 is the first generation GPS and GLONASS systems, and EGNOS is the second generation GNSS-2 system.

The Europeans intend for the GNSS-2 system to address the technical deficiencies of current systems involving systems integrity, availability, and accuracy. In addition, the who's in control issue would not only include the United States and Russia, but an international agency. An evolutionary European satellite system would eventually provide an independent navigation capability complimenting GPS and GLONASS but capable of operating without them.

6.5.3 Far East Satellite Programs

The National Space Development Agency (NASDA) of Japan is developing its own GPS technology, but it intends to link navigation and communications processes. The idea of "piggybacking" navigation payloads on telecommunications satellites has merit from a cost savings standpoint and could lead to huge numbers of satellites carrying this capability. In the near term Japan is investigating a satellite-based augmentation system (MSAS) that would be their contribution to the GNSS system and provide for air traffic management compatible with the Wide Area Augmentation System (WAAS), discussed in the next chapter.

The satellite situation in China is dependent on technology transfer issues. Since McDonnell Douglas first sold machine tools to China in 1994 that were diverted to missile development, the DoD has been concerned over China's plans to use GPS and GLONASS to improve weapons accuracy, especially for the Chinese DF-15 (M-9) missile. However, the Clinton administration, concerned over the trade deficit, has not pursued additional export controls relative to GPS except to stand by the 1992 decision not to export GPS receivers capable of operating above 60,000 ft altitude or above 1000 knots.

An important consequence of all of these spacecraft, plus their communication counterparts, is the increasing possibility of collisions in space with the associated fratricide caused by thousands of high-speed fragments in similar orbits around the Earth. This highlights the need for space traffic control. (The first satellite traffic control school opened in the United States in May 1998.) Space near Earth is no longer a limitless expanse but, like the sea, is a limited resource worthy of protection and care.

To conclude this chapter on GPS and its international cousins, it is safe to say that GPS, originally intended as a "force enhancer" for the U.S. military, may be the most cost-effective defense investment ever made. It is a great buy for America and is continued evidence of American ingenuity and innovation. GPS is the prime example of a dual-use philosophy that benefits both the civilian and military sectors of our society.

6.6 Closure

The Soviet Union launched the first satellite, Sputnik, on November 3, 1957. Although the satellite's orbit was extensively tracked, two scientists at Johns Hopkins University, William Guier and George Wieffenbach, developed an ingenious scheme to fully characterize Sputnik's orbit by observing the Doppler shift in its radio waves as it passed overhead. Other researchers realized that the scheme, if reversed, could be used to locate an Earth-based observer provided the satellite's orbit was known. Thus was born Transit, a United States Navy satellite positioning system that was used to locate ships and submarines from 1964 until Transit was retired in 1996.

Brad Parkinson is the individual who is largely responsible for defining the current GPS concept. While managing the GPS program for the Air Force in the early 1970s, he competed with the Transit program for scarce funds at a time when GPS was very vulnerable to cancellation. When the Navy desired to upgrade Transit satellites to track ballistic missiles, he convinced a senior defense official, Robert Cooper, to let GPS do the job, wresting 60 million in development funds for two spare GPS satellites, bringing the total number to six from four. This provided some security for GPS until the Gulf War and civilian innovation would irrevocably establish the concept of satellite navigation.

Parkinson, now at Stanford University, often quotes a universal law of consequence to explain how GPS was developed: "One thing leads to another." He heard the law from Robert Cannon, Chief Scientist of the Air Force, and the law continues today to explain the evolution of differential GPS (DGPS) and now carrier-phase DGPS (CDGPS). As GPS receivers add the capability to track Russian GLONASS satellites, GPS is becoming a truly international global navigation satellite system (GNSS). If communications satellites are added to the mix, a worldwide, telecommunications-navigation industry is seen to be rapidly emerging. This may impact international development in the 21st century the way electricity impacted national development in the 20th century. That is to say, "one thing will lead to another."

7
High Accuracy Navigation Using Global Positioning System

GPS is an excellent navigation system, but there are problems with signal interruptions that can have severe implications for high accuracy applications. As simple a problem as blockage of the GPS antenna can cause a serious situation depending on the task that is at hand. There are three qualities that GPS needs to become accurate enough for precision landings in aircraft and for vehicle guidance along roadways—accuracy, availability, and integrity. These concepts have mathematical definitions involving confidence levels and probability but are intuitive as well.

It was never the intent of the GPS developers for it to be a stand-alone navigation system with high accuracy applications. In fact, GPS was only intended to provide a nonprecision approach capability for aircraft by the inventors of SA. However, once the concept behind DGPS became popular and significantly improved accuracy, regardless of SA, efforts began to improve its availability and integrity as well. The original GPS concept guaranteed an availability of 21 satellites 98% of the time but had few provisions for integrity (other than the geometry of GDOP) by which users would be informed if a satellite's transmitted signal was unusable or degraded.

In this chapter the traditional, evolutionary view of achieving improved performance by combining the GPS with an inertial navigation system is presented and seen to have merit. This combination has been in various phases of successful implementation for many years. With the strong increase in system reliability and performance that has accompanied INS development in the 1980s, a significant improvement in military capability has evolved (provided that one had access to PPS with SA off). Now this evolution has been accelerated by complimentary technologies and new ideas for using GPS.

Ingenious innovations have forever changed the GPS concept. Researchers, including surveyors using GPS, realized that with a precise landmark and two GPS receivers they could eliminate nearly all of the error sources plaguing the GPS, including SA. The technique was called differential GPS or DGPS, and it exploits the fact *that selective availability degrades accuracy, but not transmitted signal precision.*

There was a strong push in the 1993 to 1996 time frame toward replacing most, if not all, of the navigation aids in the continental U.S. with a form of DGPS called the Wide Area Augmentation System (WAAS). This enthusiasm has since been tempered by inadvertent GPS signal outages and by instances of direct spoofing and jamming of the GPS signal over large areas and for long periods of time. The final section of this chapter discusses the WAAS concept and its flight applications.

After completing this chapter you should be able to do the following:

1) Describe how a GPS system that is vulnerable to line of sight and jamming may be aided by an inertial navigation system; conversely, describe how an INS

140   INTEGRATED NAVIGATION AND GUIDANCE SYSTEMS

system vulnerable to long-term drift may be aided by the GPS navigation solution; point out the implications for reliability, accuracy, and overall systems integrity.

2) Explain the fundamental concept of differential GPS (DGPS) and discriminate between local area DGPS and wide area DGPS; identify and list the significant error sources of DGPS.

3) Describe the Wide Area Augmentation Plan for implementing DGPS and explain its potential impact on military and civil aviation.

7.1 Global Positioning System Aiding Inertial Navigation System

There is a difference between using GPS as just another update device for the INS and using it as an integrated system. This difference is focused on the Kalman filter. For an integrated system the Kalman filter is not necessarily isolated within the INS. Before going into this in detail, the normal method of updating the INS will be reviewed.

7.1.1 Inertial Navigation System Update Process

One of the goals in INS testing is to determine the influence of sensor and initialization errors on system position and velocity accuracy. This is extremely difficult if a wide variety of different types of update measurements are used. A Kalman filter is a natural solution to this problem because it provides *a statistical weighting and error propagation algorithm*. Following an external "fix," the filter updates position and velocity on the basis of statistically weighting the covariance of the external measurement against the covariance of the estimated state of the system. Between updates the filter propagates estimates of the navigation solution and estimates of modeled sensors. Thus the KF is an intuitive approach that supplies an optimal estimate of navigation system errors and provides estimates of all modeled navigation sensor error sources that have significant correlation times.

In the *position update mode*, updates are made to stored "waypoints" used by the Kalman filter (KF) to compute and apply corrections to the navigation equations. Position updates reduce attitude error as well as removing position errors, thus slowing error growth between measurements. *Velocity update* obviously cannot affect position error directly, but is used by the KF to update the range rate predicted by the navigation equations and thus to prevent buildup of position errors.

A *visual flyover* accuracy for update at low level is probably accurate to within 200 ft rms. Thus, a pilot in low-level flight updates about as well as SPS GPS. At higher altitudes, of course, visual flyover updates are not accurate without instrumentation. Flyover updates must be reduced to along-track and cross-track.

Updates from forward-looking infrared (FLIR), RADAR, or other devices that employ a cursor in the cockpit are difficult to categorize relative to accuracy because of the wide range of environmental and pilot-dependent variables that contribute to it. A reasonable overall accuracy would be 25–75 m depending on the flight conditions.

GPS, when used as a source for updating, provides SPS or SA accuracy during measurement updates that are highly dependent on GDOP and UERE. From Table 6.2 in the preceding chapter, this means that SPS GPS (SA on) will provide a 20-m fix, and GPS PPS (SA off) will improve to a 5-m fix.

The results of the preceding types of integration can be seen in the familiar sawtooth update pattern from aided-INS test flights. Although performance is good for most update schemes, and excellent when the GPS is the measurement source, there is a high cost in time and effort involved with systems integration (not to mention the lack of standardization and resulting maintainability problems). The key question is how to do the integration and how much cost and effort it will require.

7.1.2 Global Positioning System Integrated with Inertial Navigation System

The primary concern with using GPS as a stand-alone source for navigation is signal interruption or blanking. As important a concern, however, is GPS systems integrity, which is the ability of an overall system to "flag" that its output has gone sour. With an INS the primary concern is long-term error growth in the navigation output. Aiding is usually done to keep this error bounded and provides the aided-INS with good system integrity.

As it turns out, GPS and INS compliment each other's weak points admirably. GPS can provide a suitably equipped user with PVT whose errors are better than most other systems, at any time and in any weather (although the error from tropospheric time delay is affected by weather fronts), even in conditions of substantial radio interference. However, it suffers from the output frequently being unavailable and delayed as compared to an INS. It also does not give estimates of acceleration or attitude rates. GPS performance, moreover, is known to degrade significantly during high rates of maneuvering. Note that a "GPS outage" in effect occurs whenever the user antenna cannot see four satellites. GPS velocity, obtained by differencing, can be very erratic during these times.

7.1.3 Global Positioning System/Inertial Navigation System Integration Architectures

Aided inertial navigation is used to keep the estimate of the "navigation state" current. External measurements are obtained and processed through an extended Kalman filter (EKF) to provide the most probable correction to the state estimate. The accuracy requirements of a modern integrated avionics system drive both hardware procurement and software development, and both must be designed to work in unison. The detailed description that explains the design is often called the *system detailed requirements document* (SDRD). The computer programs are the nerve centers of the resulting system, and the code must be made compatible with the *operational flight program* (OFP) software existing on the aircraft.

This compatibility has a severe and vital constraint—that the control of the aircraft takes priority over all other tasks. If flight control requires a 100-Hz feedback loop for stability and safety of flight, the OFP will be designed to complete the navigation solution in its spare time. The hierarchy for a computer executive routine will run vital processing for flight instruments and feedback at the kilohertz rate, the navigation and control interface and displays near 50 Hz, and update the navigation estimates near a 1-Hz rate. A computer does this by scheduling the lower rate tasks to run in the background.

The measurement updates, of course, may occur at any time. The computer must therefore time tag a measurement when it occurs and then wait in the background for the time to do the filter processing that results in new state estimates and covariances. Each level of background task has a flag that is set to prevent starting a new update cycle before the previous one completes. The flag also may be used as a time overflow indicator in the background to tell the higher priority computer tasks to skip an update cycle when required.

Although the traffic management requirements of foreground/background computer integration are severe, the scheme is generally superior to other approaches. It is important to guard against higher priority tasks changing or using vital pieces of data while the navigation algorithm waits in the background to complete its update cycle. The navigation algorithm may also interface with a human operator, as when deciding whether or not to reject an outlier measurement.

7.1.3.1 Raw data to Kalman filter architecture. Excellent performance of GPS/INS should occur when the Kalman filter accepts raw data from both the INS and the GPS rather than using the GPS only as a measurement update device. The military GPS Receiver Type IIIA is designed to interface with a five-channel GPS receiver in exactly this manner. The KF that accepts all of these inputs models a wide variety of error sources, for both the INS and GPS. New error sources in the KF from the GPS will relate to the code-loop, atmospheric path delays, time bias, etc. This filter will have over 20 modeled states. Note that in this type of scheme the GPS may not be capable of stand-alone operation.

Kalman filter in the fire control computer. The F-16 aircraft was one of the first fighters to use GPS to aid its navigation solution in the fire control computer (FCC). In this aircraft many avionics subsystems are in common with a high-speed miltiplexed serial data bus referred to as the "1553 data bus" (named after the military publication *Mil-Std-1553 serial data interface*). The FCC is the controller for all digital communications between subsystems as well as the computer processor for advanced avionics features. All HUD navigation, including air-to-air and air-to-ground combat modes (except manual bombing and snap-shoot gunnery modes), are processed in the FCC.

In this architecture for military aircraft, an aircraft's FCC has a Kalman filter that computes the most probable position along with the estimate of error corrections for the INS position, velocity, and attitude. Because of all of the communication on the 1553 bus, it may take many seconds for an estimate to be available. When the GPS is integrated into this system, it is the FCC that maintains the navigation solution updated from many other independent sources. GPS/INS integration has been thoroughly tested in many Air Force aircraft, including the F-16 and the F-111.

Kalman filters and government furnished equipment. It is common to tie reliable and proven inertial navigation systems to newly emerging GPS systems to save cost and development time. The 1553 data bus and its civilian ARINC avionics data bus equivalent are capable of doing this. In this case the IMU signals needed for feedback stabilization and inner loop control are duplicated, one set bypassing the navigation computer to perform essential stability and control and the other set sent to a KF to be combined with GPS or DGPS. This will be discussed in more detail in Sec. 7.2.1.2.

HIGH ACCURACY NAVIGATION

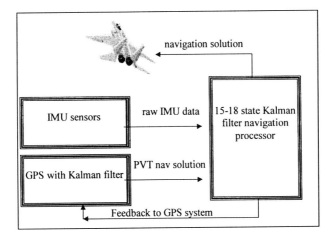

Fig. 7.1 GPSI loosely coupled system.

7.1.3.2 Loose and tight global positioning system/inertial navigation system integration architectures. In the 1980s a loosely integrated or loosely coupled configuration was developed for global positioning system/inertial navigation system integration (GPSI). In this configuration GPS position, rather than raw output, was input into the navigation computer KF. Because the GPS already had a KF of its own, this was a filter-driving-filter arrangement (see Fig. 7.1) that was susceptible to stability problems when the KF gains were changed, or "tuned," to boost performance. These gains were considered "loose," that is, they were left alone. The system's merit was that GPS could be operated in a stand-alone mode. The downside was unpredictable KF stability.

This architecture has evolved into a "tightly coupled" architecture where the navigation processor KF accepts raw inputs from both the GPS and IMU. There is no KF in the GPS processor. Unmodeled errors from the GPS KF are eliminated, and this allows the KF gains to be tuned tightly, hence the name of the architecture.

7.1.4 Embedded Global Positioning System/Inertial Navigation System

As INS and GPS technologies yield lighter and smaller navigation devices, the industry has come up with an embedded GPS/INS concept called EGI. EGI sacrifices some performance capability for low cost in addition to small size and weight. It is being tested for use on standoff weapons systems and autonomous remotely piloted aircraft of all types.

7.2 Differential Global Positioning System

Differential GPS (DGPS) corrects bias errors at one location by using a reference GPS receiver to form individual pseudorange corrections for every GPS satellite in view and then transmitting the corrections to another location. The reference

144   INTEGRATED NAVIGATION AND GUIDANCE SYSTEMS

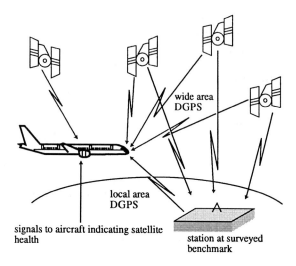

Fig. 7.2 Differential DGPS.

receiver must be at a known, surveyed spot so that it can take out all of the pseudorange errors for all of the satellites. It is a straightforward idea that has many applications because it can bring the navigation error near the 1 m range (see Fig. 7.2).

Note that this is not a simple position update that is sent from the reference receiver but an actual correction to the raw pseudoranges. A position update would have the same effect only if both receivers were assured of tracking exactly the same satellites.

DGPS is a "code-based" differencing technique and falls into two categories: local area DGPS (LADGPS) and wide area DGPS (WADGPS) or wide area augmentation. Either DGPS method eliminates all of the errors induced by SA and most of the errors induced by the atmosphere. Estimated degradation of the correction based on the difference between the two GPS systems is spatially correlated but usually not a factor for line-of-sight (LOS) operations. With DGPS the SPS navigation accuracy can be improved horizontally to better than 1 m. The corrected signal from the reference should be less than 10 s old for this to hold, and the user should be within 40 km of the reference station (accuracy degradation is insensitive to range even beyond this distance, however).

7.2.1.1 Local area differential global positioning system errors. Local area DGPS drives the following errors (meters rms) to zero: satellite clock stability (3 m), satellite perturbations (1 m), SA (32.3 m), ephemeris (4.2 m), ionospheric delay (5 m), and tropospheric delay (1.5 m). Only the receiver noise and multipath errors remain (3.3 m rms). The numerical values depend on the receiver being close to the reference, but the degradation of accuracy with distance from the reference is graceful.

A comment is necessary about tropospheric delay. This delay, more correctly called *neutral* delay, lumps *nondispersive* delays (delays that depend on the atmospheric water content or humidity) resulting from the signal passing through both the stratosphere and troposphere. Because the ionosphere is a *dispersive* weather medium, which means the delay depends on radio frequency, the ionospheric part of the total delay can be eliminated with dual-frequency receivers. The tropospheric delay, however, is variable with weather fronts and may be a problem if the baseline for DGPS exceeds 10 n miles. The worst delays from the troposphere can be avoided by elevation masking the GPS receivers, thereby avoiding signals from low elevation satellites.

If only a position error is transmitted by the reference, and not an actual satellite-dependent pseudorange correction, then it is assumed that the reference and receiver share "common" errors, which means that they must be locked-on and navigating with respect to the same SVs. Because this is impractical to ensure (if more than five satellites are LOS the possible combinations are very high), the pseudorange solution is much preferred.

The transmission characteristics of the message from the reference are being standardized. This will allow use of DGPS at airports for ground control as well as for air traffic. It is also important that the same satellite ephemeris data be used by both GPS systems. This will take some prior planning since they are periodically revised. Finally, the GPS receivers should use digital signal processing to lower the noise floor of the receiver because this is now the greatest source of error in DGPS.

If a third civil signal is introduced into GPS satellites launched in year 2004 and thereafter, a technique called *wide laning*, which measures the differences between frequencies when performing DGPS calculations, would be greatly simplified. This procedure (not explained here) could increase the distance between a user and a reference station over which DGPS accuracy is maintained by an order of magnitude.

A final point about DGPS and SA. DGPS defeats SA, but it does so only if the measurement correction is current. If the correction is delayed by more than a few seconds in the user's computer, then SA as the dominant error source will have reintroduced error. Tests have shown that the initial growth of this type of error grows as time squared and that the correction from the reference should be used within 10 s from transmission.

7.2.1.2 Commercial inertial navigation system/global positioning system. The military recognized immediately that the INS could help the GPS remain pointed to satellites during dynamic maneuvering and for brief signal outages. Civilian aircraft, however, could not enjoy this capability because SA caused difficult tracking problems for the GPS receivers. In engineering terms, the GPS code and carrier tracking loop bandwidth could not be reduced enough to track in the presence of SA because the penalty was reduced radio interference rejection, which is required because of the low spread-spectrum signal coming into the GPS receiver.

In military aircraft the GPS is used to update and correct for INS error propagation, promising to provide a true "inflight alignment" capability while maneuvering. In civilian applications, however, the SA dithering produced GPS updates that were "correlated" with each other, and INS aiding was not practical without independent measurements. Attempts to model SA using autoregressive models have shown promise in successfully predicting SA many seconds into the future, but it was the advent of DGPS that opened commercial aircraft to the benefits of combined GPS/INS.

As initially described in Sec. 7.1.3.1, when two existing INS and GPS systems are integrated, the signals from the INS required for feedback stabilization are often split, one set going to the flight stabilization algorithm and the other to the navigation algorithm to be mixed with GPS.

The GPS data and DGPS transmitted corrections are available near a 1-Hz rate. If the system requirements include automatically landing the aircraft, the 1-s delay would be intolerable and unsafe. This problem is usually solved by using a DGPS-corrected measurement from the past to update the KF states at some time in the past, and then by propagating the state estimate to the current real time when flight controls are being applied. This technique allows continued use of automatic guidance and control in the event of GPS or DGPS transmission link dropouts. For this type of systems integration, the baro-altimeter remains a key provider of external measurements in every phase of the approach and landing.

Another important consideration is the difference between GPS time and IMU time. This difference must be determined from respective time-tags when they are available and then interpolated to real time. Computer *buffers* or *registers* hold these time-tags along with GPS pseudorange data until DGPS corrections are available. At update time (a known time-tag in the past), the corrected pseudoranges and the baro-altimeter measurements are used to correct the navigation state propagated to update time by the KF. The corrected state is then propagated to real time and used for stabilization, control, and operator displays. This estimate continues to propagate and be available as needed until it is intercepted by the next correction resulting from an update in the past. The accuracy of this type of system is 1–3 m CEP.

7.2.2 Nationwide Differential Global Positioning System Network

The Nationwide Differential GPS (NDGPS) Network started on the East and West coasts as reference stations operated by the U.S. Coast Guard, but now covers a substantial part of the nation, including the eastern third of the country. This network, established by legislation in the Fall of 1997, is fully integrated with the U.S. Coast Guard's radiobeacon system and is jointly run by the Coast Guard and the Army Corps of Engineers. It is also compatible with international DGPS systems that eventually should provide a seamless worldwide navigation system.

Expansion plans were boosted by a U.S. Air Force decision to decommission the Ground Wave Emergency Network (Gwen), a casualty of the end of the Cold War that freed up 22 Gwen sites for conversion to NDGPS. At the time of this writing, it is unclear how the NDGPS will impact aviation. NDGPS goals include system *availability* 99.999% of the time, *accuracy* from 1–5 m CEP, and system *integrity* by providing an alarm when a satellite signal becomes unreliable.

7.3 Wide Area Augmentation

As the baseline distance increases greatly (above 100 km) between a reference and user GPS in a DGPS arrangement, the accuracy of DGPS will degrade significantly. The passage of a significant weather front will cause additional degradation from additional tropospheric delay. Perhaps a thousand transmitting references would be required to cover the continental U.S. with broadcast pseudorange corrections. WADGPS provides LADGPS accuracy over a wide area with only a handful of reference stations. The idea is to make the navigation correction signal an estimate valid for a large area, where the estimate is allowed to degrade at the perimeters of the area. It thus would behave analogously to a conventional radionavigation aid.

The WADGPS master station for a large region (20–30 planned for the continental U.S.) would receive DGPS correction data from reference monitor stations placed to have overlapping LOS to the GPS satellites. The master station computer estimates the error correction vectors for selected large regions where aircraft are operating. These corrections are sent to GPS satellites, which broadcast the corrections to the user as determined by the user's general location.

The Federal Aviation Administration has embraced this idea: "The FAA is working with industry to demonstrate the feasibility of transmitting extremely accurate, local-area, differential GPS corrections to aircraft within airport-line-of-sight distances to provide all-weather approach capability at any runway within the coverage area. Local Area Augmentation Systems (LAAS) from several contractors are expected to undergo flight readiness review and flight testing." This application of WADGPS is called the Wide Area Augmentation System (WAAS; see Fig. 7.3), and WADGPS can be considered the basis for WAAS. When functional and tested, WAAS will allow the "free flight" concept to be seriously considered nationwide. This allows pilots to choose optimal routing by their own criteria.

An error budget for WADGPS shows that ephemeris (0.4 m), tropospheric (0.3 m), and ionospheric errors (0.5 m) dominate the RSS error, but the navigation accuracy with HDOP of 1.5 is 1.2 m. This assumes that the information is current and that the user has a dual-frequency unit to estimate ionospheric time delay. If the correction information is old, tests show that 5-m accuracy can be expected after 20 s.

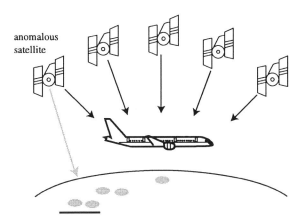

Fig. 7.3 Wide area augmentation system.

7.3.1 Ground Proximity Warning Systems and Global Positioning System

The WAAS promises to provide a precision approach capability at a host of new locations around the nation. This must occur, however, without compromising flight safety. Controlled flight into terrain (CFIT) is by far the largest overall safety-of-flight threat and is responsible, according to the Air Safety Foundation, for three-fourths of the total fatal turbine aircraft accident risk. Because of this, ground proximity warning systems (GPWS) are now required on all commercial turbine aircraft with 10 or more passenger seats. The problem with GPWS is the short warning time it provides (typically 10–15 s before impact) and the resulting situational awareness and reaction demands on the pilot. Moreover, when an aircraft is fully configured for landing, the system shuts itself down.

When GPWS is combined with GPS and a worldwide terrain database, however, warning times can be greatly increased—to as much as a full minute in certain modes of flight. On approaches the combined system builds a terrain clearance floor of 500 ft at the runway that expands to 800 ft up to 25 miles away. Penetrating this floor brings an aural warning from the system. By comparing GPS position to the database, the system can continue operating while the aircraft is fully configured to land, as is the case for an aircraft maneuvering during a nonprecision approach.

7.3.2 National Satellite Test Bed

The National Satellite Test Bed (NSTB) is a prototype WAAS operated for research purposes by Stanford University. Recall that local area DGPS transmits a *scalar* pseudorange correction in which all errors are significantly reduced simultaneously. The NSTB provides a *vector* correction of errors modeled separately to aircraft in various phases of flight. Users then spatially decorrelate the errors that are most important at their location.

NSTB has 29 dual-frequency reference stations with a nominal separation of 500 km. Error models are developed based on reference stations' observations for a satellite's clock, ephemeris, and ionospheric delay, plus local models for receiver noise and multipath. These models provide GPS error terms over continental reference scales that are broadcast by a geosynchronous satellite link to users. The received navigation solution accuracy is approximately 2 m CEP.

The test bed addresses the three characteristics lacking in conventional GPS for flight purposes: accuracy, integrity, and availability. These characteristics will be more extensively discussed in the next chapter.

7.4 Closure

GPS was developed by the U.S. Air Force as a joint program with other military services. In the development process there were many competitions solicited among electronics suppliers [these were called nondevelopmental item (NDI) competitions]. These competitions resulted in the precision lightweight GPS (PLGR) receiver, the miniature airborne (MAGR) GPS, and the embedded (EGI) GPS. These items are now renewing our military navigation capability, our weapons delivery methodologies, and our test range operating procedures.

By the year 2000, nearly 100,000 lightweight receivers, 10,000 airborne units, and 6000 weapons GPS receivers will be in active military service, and the GPS industry will be a $30 billion per year operation. Congress, NASA, and the FAA have shown intense interest in GPS development and deployment in systems they are developing for both land-based and space-based navigation. Television coverage of the Gulf War (when selective availability was turned off) generated high public interest in GPS that continues unabated today.

In many ways aerospace has not received the press it deserves on cost. General Motors spent $3.6 billion on the Saturn automobile. The Pentagon spent $2.6 billion creating the stealth fighter using F-16 flight controls, F-18 engines, and off-the-shelf avionics that can cost $7000 per pound in a new aircraft. The F-22 will have the equivalent of seven 1995 Cray supercomputers on board (truly a flying "dust cover" for electronics). GPS will overshadow all of these programs in expenditures over time. Like the internet, GPS represents the enormous potential synergy that can occur when civilian innovation is allowed to capitalize on research investments by government and the military.

8
Differential and Carrier Tracking Global Positioning System Applications

Applications of the carrier-based techniques for rapid, precise navigation are now fully appreciated. This chapter is devoted to describing a few of the more salient and promising applications that may have major impact on flight operations and on the aerospace industry as a whole. They also have vital implications for the national economy, and there seems to be no limit to new and useful innovations that are constantly occurring. Aviation has also been adapting rapidly to exploit these emerging developments. The overriding purpose of this chapter is to *provide you with an understanding of what is involved in applying differential and kinematic, carrier-phase GPS to practical aircraft applications.*

After you have finished this chapter you will be able to do the following:

1) Understand the underlying principles of differential carrier-phase tracking.

2) Describe the formidable integrity problems that must be solved to allow CAT III aircraft approaches using satellite navigation systems and how they are addressed.

3) Explain how kinematic GPS can be used for attitude control of aerospace vehicles.

4) Survey the far-reaching implications of using satellite navigation systems for aircraft traffic control, including the free-flight concept.

5) List the applications of GPS systems to test ranges, including the impact on time synchronization and measurement.

8.1   Review of the Kinematic Carrier-Phase Methodology

Advanced GPS receivers are capable, once locked onto the L_1 and L_2 carrier signals of a satellite, of keeping a cumulative cycle count of the Doppler offset frequencies caused by relative satellite motion. This count begins from a sample data set taken from the first measurement epoch of active tracking at some unknown integer value N, which, if determined, precisely places the satellite at a distance of $N\lambda$ from the user at start of track, where λ is the signal wavelength (19.03 cm for the L1 carrier and 24.42 cm for the L2 carrier). Moreover, using interferometric techniques the fraction of the wavelength between epochs may also be measured by capable GPS receivers to within 1% of the wavelength, giving rise to extreme positioning accuracies in the millimeter range if the $N\lambda$ ambiguity problem can be precisely solved. These accuracies break through the receiver's *noise floor*, the largest error source for DGPS, and take advantage of the accuracies inherent in the transmitted signal.

Surveyors using post processing of data have achieved millimeter accuracy from GPS field measurements of fixed positions on the Earth. Users who are moving relative to the Earth's surface must solve the $N\lambda$ ambiguity problem "on the fly" and have a more difficult problem. For aircraft the $N\lambda$ ambiguity problem is typically solved by determining the "baseline" distance between a known surveyed position on the ground and the GPS antenna on a moving aircraft. This is called *differential carrier-phase tracking* (CDGPS) or *real-time kinematic* (RTK) GPS. The process uses a Kalman filter and a double differencing (DD) technique to obtain a solution to the integer ambiguity problem. The length of time it takes to obtain a solution is related to the change of geometry that must occur between the satellites and the user to resolve the integer ambiguity. This time can be reduced using pseudolites on the ground.

Pseudolites, or integrity beacons, are ground transmitters on the GPS L_1 signal frequency that are modulated with their own pseudorandom code (PRN) so that they are not mistaken as orbiting SVs. Presuming a rapid change in geometry between user and the ground-based pseudolite, the integer ambiguity problem can be solved in a few seconds if the user can receive, process, and integrate the pseudolite signal with the onboard GPS processor. Centimeter accuracies have been demonstrated with this procedure. Use of these devices significantly enhances the accuracy, availability, and integrity of GPS.

8.1.1 Introduction to the Carrier Modeling Process

The purpose of this section is to provide an introduction to the double-differencing technique for kinematic carrier-phase modeling. The process has been simplified to its essentials so that the underlying principles are emphasized. The scenario considers two fixed antenna locations, user receivers A and B, and uses carrier-phase modeling to determine the line-of-sight (LOS) distance r_{A2B} within a fraction of the wavelength ($\lambda_{L1} = 19$ cm) of the L1 carrier. This L1 carrier is reproduced in all user receivers, considered here to have clocks that can generate L1 internally with no errors.

Figure 8.1 illustrates the situation for fixed locations A and B when the receivers are locked onto space vehicle #6 (SV6). Focusing on the user at receiver A for the

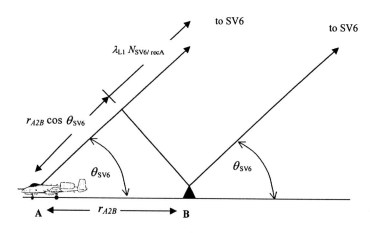

Fig. 8.1   CADGPS geometry.

DIFFERENTIAL AND CARRIER TRACKING APPLICATIONS

moment and realizing that the LOS angles to SV6 from either user A or user B are for all practical purposes the same, the following equation may be written:

$$r_{SV6/recA} - r_{SV6/recB} = r_{A2B} \cos\theta_{SV6} = r_{A2B}(\mathbf{i}_{SV6/recA} \cdot \mathbf{i}_{A2B}) \quad (8.1)$$

The goal is to find a solution for distance r_{A2B} from differential phase measurements at user receivers A and B. Assuming for illustration purposes that the speed of light c is absolute and that there are no error sources to compound the situation, the distances from the SV6 antenna to the user antennas are given by

$$r_{recA/SV6} = c(t_A - t_{SV6\_send})$$

$$r_{recB/SV6} = c(t_B - t_{SV6\_send}) \quad (8.2)$$

where t_A is the GPS time that antenna A received the phase signal sent by SV6 at t_{SV6_send} and t_B the time user B receives it. It will be shown that the solutions for r_{A2B} involve DD operations on the receiver A and B carrier-phase measurements.

To relate relative distance to received phase difference, look at the L1 signal in Fig. 8.2, where the x axis is scaled to show fractions of an L1 cycle. Because the speed of light is $c = f_{L1}\lambda_{L1}$ (frequency times wavelength), the x-axis variable is equivalent to

$$\frac{\omega_{L1} t}{2\pi} = f_{L1} t = \frac{c}{\lambda_{L1}} t = \frac{r_{SV6/recA}}{\lambda_{L1}}$$

$$\lambda_{L1} \cong 19 \text{ cm for L1 carrier} \quad (8.3)$$

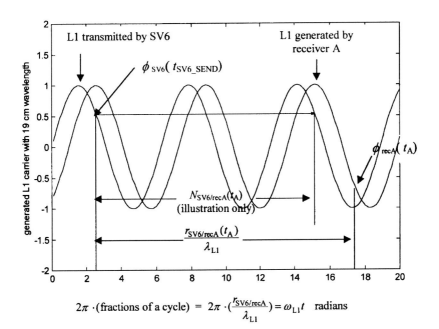

Fig. 8.2 L1 carrier received phase difference.

154 **INTEGRATED NAVIGATION AND GUIDANCE SYSTEMS**

The "perfect" receiver at user antenna A measures a phase difference at time t_A given by

$$\phi_{SV6/recA}(t_A) = \phi_{SV6}(t_{SV6\_send}) - \phi_{recA}(t_A) \tag{8.4}$$

but from Fig. 8.2, with phase ϕ expressed in fractions of an L1 cycle,

$$\phi_{SV6}(t_{SV6\_send}) = \phi_{SV6/recA}(t_A) - \frac{r_{SV6/recA}}{\lambda_{L1}} + N_{sv6/recA}(t_A) \tag{8.5}$$

where $N_{sv6/recA}(t_A)$ is an unknown integer multiple of wavelength λ_{L1} at time t_A. Substituting Eq. (8.5) into Eq. (8.4),

$$\phi_{SV6/recA}(t_A) = \phi_{SV6}(t_A) - \frac{r_{SV6/recA}}{\lambda_{L1}} + N_{sv6/recA} - \phi_{recA}(t_A) \tag{8.6}$$

For the user receiver at B that is taking simultaneous phase measurements from SV6,

$$\phi_{SV6/recB}(t_A) = \phi_{SV6}(t_A) - \frac{r_{SV6/recB}}{\lambda_{L1}} + N_{sv6/recB} - \phi_{recB}(t_A) \tag{8.7}$$

where it is assumed in Eq. (8.7) that both user receivers at A and B sample the phase-tracking signals at exactly the same time t_A.

The relative distance r_{A2B} now can be found by differencing Eqs. (8.6) and (8.7) and using Eq. (8.1) in the result:

$$\Delta\phi_{SV6}(t_A) = \phi_{SV6/recA}(t_A) - \phi_{SV6/recB}(t_A)$$

$$= -\left[\frac{r_{A2B} \cos \theta_{SV6}(t_A)}{\lambda_{L1}}\right] + [N_{sv6/recA}(t_A) - N_{sv6/recB}(t_A)]$$

$$- [\phi_{recA}(t_A) - \phi_{recB}(t_A)] \tag{8.8}$$

If another satellite SV18 is available, the preceding process may be duplicated so that

$$\Delta\phi_{SV18}(t_A) = -\left[\frac{r_{A2B} \cos \theta_{SV18}(t_A)}{\lambda_{L1}}\right]$$

$$+ [N_{sv18/recA}(t_A) - N_{sv18/recB}(t_A)] - [\phi_{recA}(t_A) - \phi_{recB}(t_A)] \tag{8.9}$$

A DD now may be done to eliminate the phase difference terms for users A and B at the far right of Eqs. (8.8) and (8.9) and produce the following vector equation:

$$\Delta^2 \phi_{(SV6\text{-}SV18)} = \Delta\phi_{SV6}(t_A) - \Delta\phi_{SV18}(t_A)$$

$$= -\left[\frac{\cos \theta_{SV6}(t_A) - \cos \theta_{SV18}(t_A)}{\lambda_{L1}}\right] r_{A2B} + \Delta N_{SV6\text{-}SV18}(t_A)$$

$$= \left[-\left[\frac{\cos \theta_{sv6}(t_A) - \cos \theta_{sv18}(t_A)}{\lambda_{L1}}\right] \quad 1\right] \begin{bmatrix} r_{A2B} \\ \Delta N_{SV6\text{-}SV18}(t_A) \end{bmatrix} \tag{8.10}$$

where the unknown integer $\Delta N_{\text{SV6-SV18}}(t_A)$ is a difference of integers given by

$$\Delta N_{\text{SV6-SV18}}(t_A)$$
$$= [N_{\text{sv6/recA}}(t_A) - N_{\text{sv6/recB}}(t_A)] - [N_{\text{sv18/recA}}(t_A) - N_{\text{sv18/recB}}(t_A)] \quad (8.11)$$

Another equation may be added to the vector equation (8.10) by waiting for time τ to pass so that $t = t_A + \tau$ results in new LOS angles. Because the locations are fixed, r_{A2B} will not change. The integer counts given in Eq. (8.11) will change by a known amount if satellite lock is not lost so that

$$\Delta N_{\text{SV6-SV18}}(t_A + \tau) = \Delta N_{\text{SV6-SV18}}(t_A) + N_{\text{KNOWN}} \quad (8.12)$$

This allows N_{KNOWN} to be brought to the left-hand side at $t = t_A + \tau$ so that the resulting vector equation becomes solvable:

$$\begin{bmatrix} \Delta^2 \phi_{(\text{SV6-SV18})}(t_A) \\ \Delta^2 \phi_{(\text{SV6-SV18})}(t_A + \tau) - N_{\text{KNOWN}} \end{bmatrix} = z = Hx$$

$$= \begin{bmatrix} -\left[\dfrac{\cos\theta_{\text{sv6}}(t_A) - \cos\theta_{\text{sv18}}(t_A)}{\lambda_{L1}}\right] & 1 \\ -\left[\dfrac{\cos\theta_{\text{sv6}}(t_A+\tau) - \cos\theta_{\text{sv18}}(t_A+\tau)}{\lambda_{L1}}\right] & 1 \end{bmatrix} \begin{bmatrix} r_{\text{A2B}} \\ \Delta N_{\text{SV6-SV18}}(t_A) \end{bmatrix}$$
$$(8.13)$$

An additional satellite will add more equations than unknowns to Eq. (8.13) and result in the familiar form for a least-squares solution for the right-hand side:

$$\hat{x} = [H^T H]^{-1} H^T z \quad (8.14)$$

Although the preceding development is greatly simplified because no errors were presumed in the measurements, it may readily be observed that the change in geometry required between satellite and user to produce linearly independent rows of the H matrix will take some time. Considering the form of Eq. (8.14), a measure based on the trace of $[H^T H]^{-1}$ similar to GDOP may be defined that reflects confidence in the solution.

If one of the users is in motion with respect to the Earth at lock-on, the problem is much more difficult than presented here. In this case a ground-based GPS satellite (pseudolite) may be implemented to produce a rapid change in geometry. Once the solution converges, solutions in the centimeter range may be expected until loss of lock.

8.1.2 Global Positioning System Ambiguity Resolution Techniques for Fixed Users

From the preceding introduction, the basic *carrier-phase measurement*, or *observable*, is the phase difference (instantaneous fractional phase) between the locally generated L1 carrier and the incoming L1 carrier from a satellite. From this measurement the fractional L1 cycle between the SV and user antennas can be determined. The number of whole cycles as counted from the initial measurement time is also available. What is missing is the *initial number* of whole L1 cycles between SV and user at the initial measurement time.

This ambiguity resolution (AR) problem may be solved by using DD to solve for a real number approximation to an unknown integer difference or by triple-differencing to remove all ambiguity terms from the derived equations. Both techniques rely on a change in user-to-satellite geometry, and thus GDOP will influence the solution.

If the users at A and B are fixed with short baseline distances between them, then an antenna swap back and forth while both users maintain lock onto the same satellites will allow data reduction techniques to solve the AR problem to millimeter precision. A change of geometry is not required for the antenna swap procedure. A variation of this procedure, in which only one user receiver takes data at two surveyed locations, or twice at the same surveyed location, also will solve the AR problem. For this case a large change of geometry is not only not required, it is not allowed.

An interesting variation of this problem is to solve for integer linear combinations of L1 and L2 carrier cycles where the wavelength of the linear combination is relatively long. Some of these linear combinations are named *wide lane* (86 cm), *narrow lane* (11 cm), and *extra-wide lane* (163 cm). If dual-frequency receivers are available that are capable of precise measurements, instantaneous ambiguity resolution is possible from a single epoch of data. The combination of GPS and GLONASS and the addition of a new civilian frequency will enhance these techniques for stationary users.

8.1.3 Global Positioning System Ambiguity Resolution Techniques for Moving Users

On-the-fly AR techniques are complex and converge to a solution in a few seconds given a geometry change between user and satellite. A practical way to bring this about is to use ground-based GPS satellites, or pseudolites, that the aircraft overflies to cause the geometry change. The mathematics underlying this technique are beyond the scope of this work, but an excellent description of the process may be found in Pervan and Parkinson (1997). An important constraint in using on-the-fly algorithms is that two pseudolites are needed for a solution "along and cross-track." For aircraft a baro-altimeter source for altitude is also highly recommended.

8.2 Aircraft Precision Approach and Landing Using Differential Global Positioning Systems

The advertised accuracy (2 drms = 100 m) of commercial SPS GPS meets nonprecision approach criteria. GPS developers until the early 1990s had no thought of using GPS for CAT I precision approaches. GPS not only was not accurate enough horizontally, but it had a VDOP/HDOP ratio of 1.4, a slow update rate, and insufficient integrity. DGPS, however, had by the mid 1990s shown

DIFFERENTIAL AND CARRIER TRACKING APPLICATIONS

Table 8.1 Summary of precision approach minimums

Category	Runway visual range (RVR)/visibility	DH, ft
CAT I	2400 ft/0.5 mile	200
CAT II	1200 ft/NA	100
CAT IIIa	>700 ft/NA	<100
CAT IIIb	150 ft < RVR < 700 ft	<50
CAT IIIc	RVR < 150 ft	0

that it is possible to *make precision approaches anywhere* within a very expansive area with *no navigation aid at the landing site* and no transition required to a final segment (because DGPS accuracy is insensitive to distance from touchdown). CDGPS, moreover, has demonstrated in flight tests the sensor accuracy required for CAT III approaches [decision height (DH) = 50 ft]. These same systems have the potential to be used for collision avoidance, traffic alert, and for guidance after landing to the final parking spot.

However, DGPS must meet stringent availability, accuracy, and integrity requirements if it is to be used as a precision landing system. Until 1994 DGPS-guided flight was never thought capable of meeting these requirements, but two methodologies ("tunnel approach specifications" and CDGPS) for enhancing GPS overall system capability are advancing rapidly; CAT I precision approaches using DGPS (aided by an INS and baro-altimeter) have been demonstrated in research programs. The requirements for a precision approach and proposed GPS criteria are reviewed in the next sections. The criteria are based on the precision approach minimums as shown in Table 8.1.

The criteria for GPS CAT I, II, or III landing is not yet set. But based on past requirements and current FAA practice, they are estimated to look like the set in Table 8.2, where only CAT IIIb is considered.

8.2.1 Accuracy

Accuracy as defined by the FAA for aviation purposes is most often expressed as 2 drms (twice the rms radial distance error). As defined in Sec. 4.2.2, in one

Table 8.2 Estimated CAT IIIb DGPS criteria 2 drms

Criteria	Horizontal requirement	Vertical requirement
Sensor error	4 m	0.6 m
Course centering error	3.5 m	0.5 m
Control error	3.2 m	0.3 m
Protection level	6.1 m	2.3 m
Max signal duration out of protection level	2 s max	2 s max
Probability of undetected error per landing	$0.5(10)^{-9}$	$0.5(10)^{-9}$

dimension 2 drms is the 95th percentile, or $2\sigma_h$, for vertical altitude error. Two-dimensional 2 drms for the horizontal plane is the radius drawn from the point of no error containing 98% (95%) of all Gaussian position data points depending upon the circular (elliptical) bivariate data distribution. For GPS this implies 2 drms= $[\sigma_r \cdot \text{HDOP}]$ for horizontal accuracy (use 2 drms= $[\sigma_r \cdot \text{VDOP}]$ for vertical accuracy).

Accuracy (or its inverse, error) is divided into three parts for piloted flight: sensor error (the difference between the estimated and true navigation state), course centering error (the difference between is computed and what is displayed), and flight technical error [(FTE) is the difference between the estimated and desired navigation state]. FTE may be considered the result of human pilot control. The *total system error* is the RSS of the sensor, course-centering, and FTE errors.

The sensor accuracy for precision approach is based on approach categories CAT I, CAT II, and CAT IIIabc, in turn based on visibility or runway visual range (RVR) and decision height (DH). CAT III, in addition, has stringent requirements for equipment redundancies. The sensor accuracy requirements are an order of magnitude greater for a precision approach over that required for a nonprecision approach. The anticipated DGPS criteria for a CAT IIIb approach is shown in Table 8.2. Although DGPS accuracy meets the horizontal requirement, it does not meet the vertical one if unaided.

8.2.2 Integrity

Integrity in aviation terms implies that a system and its monitors will somehow "flag" the normally useful output when it should not be used operationally (note that the monitors need not be on the aircraft). As of 1999, *GPS and its monitors do not have the integrity needed for flight as a stand-alone system.* Breakdowns can take exceedingly long to be recognized depending on where and when they occur. An example of excellent systems integrity is the traditional air traffic control use of surveillance (radar) that is independent from radionavigation, providing exceptionally high overall system integrity. When a single system is modified to meet integrity issues, it may fail to meet availability requirements, that is, the percentage of time which it must be fully operational.

A disaster may occur when a degradation, or "signal out of protection level," occurs in a precision approach aid and the degraded or failed status is not flagged to the pilot. This is called *missed detection probability* and for approaches is in the 10^{-9} area (see the last two rows in Table 8.2). DGPS has three ways to achieve this extremely low value: 1) add an integrity monitoring channel that is controlled externally, 2) implement RAIM, 3) implement extra ground pseudolites for error checking.

RAIM involves integrity management by internal checks and balances. RAIM without aiding from another source implies at least five satellites in view (and usually six) to meet the expected DGPS criteria for a CAT I approach. This criteria would make the probability 0.999 that a stand-alone DGPS could finish an approach to precision minimums at the expected accuracies, after losing the most critical satellite in view. It has been tested and shown that DGPS requires external source

DIFFERENTIAL AND CARRIER TRACKING APPLICATIONS 159

aiding to meet this requirement (baroaltitude aiding allows DGPS to meet CAT 1 requirements).

8.2.3 Availability

The percentage of time that GPS must be useable (GDOP < 3) for required navigation performance (RNP) is estimated to be set at 0.99999 or $(1 - 10^{-6})$. For commercial flights GPS may have to demonstrate this availability with the loss of a most critical satellite. For DGPS the ground reference, external monitor, and the SV signals would all have to meet the availability requirement. Because it is nearly impossible to meet this with the current constellation mix, DGPS will require aiding of some sort. DGPS aided with INS, a baroaltimeter, and pseudolites is able to meet the availability requirements for CAT I approaches.

8.2.4 Tunnel Concept

The total system error, introduced in Sec. 8.2.1, is the RSS of the sensor, course-centering, and flight technical errors. There is a strong investigation by the FAA in what is called the *tunnel concept* for precision approach, a concept that focuses on total system error rather than sensor error when specifying RNP. The tunnel is a tube in space that the aircraft must stay within during approach to landing. The aircraft must remain in an inner tunnel with a probability of 95% and within an outer tunnel with a probability of $[1 - (10)^{-7}]$.

Instead of specifying the performance of each component, the performance of the resulting system is specified. This is not a redefining of the component specifications to allow lower standards but a recognition that a tradeoff is allowed between sensor error and flight technical error. The tunnel dimensions decrease as distance decreases to the runway, and it is up to flight test and simulation to verify that the probabilities for penetration of the inner and outer tunnel walls are met. This would result in landing specifications as in Table 8.1, which vary with the estimated accuracy of the sensors and autopilot.

8.2.5 Precision Approach Using Pseudolites

The preceding discussion shows that it is exceedingly difficult for GPS to replace precision aids for landing, and that even if accuracy is obtained, the redundancy and monitoring required by integrity considerations remain an overwhelming challenge. A team at Stanford University, however, demonstrated and exceeded all of the estimated specifications for accuracy, availability, and integrity that apply to precision landings in an actual flight test of a Boeing 737. Their work was so impressive that Daniel Goldin, the Head of NASA, and David Hinson, the Head of the FAA, witnessed and participated in some of the flight tests in October 1994. The system was called the *integrity beacon landing system* (IBLS). This system was the first time that pseudolites were used to resolve the integer ambiguity problem.

The basic concept is shown in Fig. 8.3 and is based on low-power pseudolites on the ground that mimic GPS satellites. These pseudolites (the name originated from psuedo-satellite) are renamed *integrity beacons* to emphasize their integrity function. They are put in pairs for redundancy on each side of the runway and

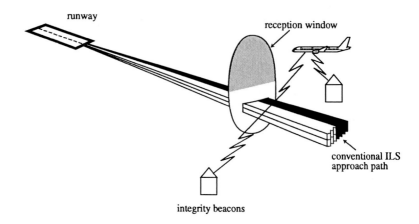

Fig. 8.3 Precision landing with IBLS.

create a window of reception along the final approach path that allows precise positioning both along and cross-track. The aircraft receives the signal when in the window and resolves the integer ambiguity problem that is associated with CDGPS in a few seconds. The aircraft navigation system now has *centimeter level accuracy* and will maintain this accuracy level as long as it is locked onto the same satellites and is receiving data from the ground station. This level of accuracy may thus be maintained throughout taxi operations on the ground.

The system flight demonstration on a Boeing 737 in 1994 made 110 out of 111 successful automatic landings to touchdown using IBLS guidance. Touchdown accuracy was only limited by the accuracy attainable by the autopilot and not by the guidance command. The one approach that was not carried to touchdown was called off by the IBLS because it detected an integrity flaw. As it turned out, a U.S. Air Force data upload to one of the satellites being tracked made the satellite inoperative for a moment. The IBLS flagged the problem, and the pilot discontinued the approach.

The integrity beacon is an ingenious solution to the precision approach problem. The beacon may be counted a sixth satellite for availability purposes (pending FAA approval). Because there are two of them, if one fails there is still 0.999 probability of making the approach to precision minimums.

8.3 Aircraft and Spacecraft Attitude Control

In 1993 the Air Force RADCAL satellite demonstrated accurate spacecraft attitude control using two cross-strapped GPS receivers. The receivers were set up to make differential carrier-phase measurements. The demonstrated capability will benefit low-earth-orbit missions where such control is critical.

In the same time frame, flights of a NASA King Air using a Trimble receiver demonstrated attitude control accuracy of 0.05 deg one sigma. Antennas were placed at various points on the aircraft to resolve the integer ambiguity, and the system had no problem tracking new satellites at the required resolution for attitude determination as they came into view. Roll angles were up to 60 deg of bank for the tests. The technology for attitude determination is promising, certainly for low dynamics transport aircraft.

8.4 Air Traffic Control

Today controlled airspace, military or commercial, is defined by navigation aid or radar fix. Area navigation (RNAV) allows the use of three-dimensional waypoints and lets users choose their route using these waypoints. The system, however, is not expected to handle the predicted numbers of aircraft that will soon be entering it.

GPS provides worldwide navigation capability to fly preferred routes, approach system capability to nearly any location, and reduced separation in nonradar airspace in conjunction with data links. The government as it transitions to GPS may be able to reduce its vast infrastructure of transmitters that provide navigation services.

8.4.1 Local Area and Wide-Area Differential Global Positioning Systems

FAA flight tests have demonstrated that CAT I sensor accuracies are met in LADGPS tests when the VDOP is low (less than 2). Lateral accuracy requirements for CAT III approaches have also been verified in flight. Wide-area DGPS is still in the concept phase, but analysis indicates that vertical sensor accuracies will be a problem for CAT I approaches. It may be necessary to provide only this level of precision approach accuracy at major airports (with integrity beacons for example), with a reduced but still significant precision approach capability elsewhere.

Researchers are studying long-range (250 n miles) RTK GPS that, if proven feasible, may provide centimeter-level accuracy over large areas of the nation. Based on the creative talent of these individuals and their track record of innovation, the idea of centimeter-level accuracy nationwide should not be dismissed as hopelessly visionary.

8.4.1.1 Approach and traffic control.
DGPS has been shown to be clearly superior to current landing approach aids in increasing the traffic flow for parallel runways. It also reduces the obstacle clearance area required near airports, allows any type of curved approach, and has no problem operating in remote areas. DGPS in analysis has shown a long-term four-dimensional capability to predict separation conflicts that is equal to the short-term prediction capability today.

8.4.1.2 Reduction of congestion in airspace.
The four-dimensional (position plus time) navigational capability that GPS technology has made possible will significantly enhance the long-range conflict prediction capability of air traffic control. Moreover, the use of curved approach paths and closer separations on parallel runways will promote safer operation in the congested space near airports. RNAV flights should be the norm under GPS surveillance rather than the exception.

Flight management systems can be programmed to provide four-dimensional guidance, but the air traffic control system should allow and encourage it under GPS because the navigation accuracies as reported by GPS do not depend on location or angular deviations from a transmitter. Increased aircraft surveillance is possible via the Mode S transponder on commercial aircraft, which is capable of transmitting GPS aircraft digital data to surrounding aircraft as well as air traffic control.

8.5   Test Ranges

The first DGPS test by the NAVSTAR office was at Yuma in 1979 as an evaluation of its suitability as a reference device. It resulted in GPS receivers (TI-4100), some of which were used by the B-2 Combined Test Force (CTF) at Edwards Air Force Base, California, in 1987 to generate time-space-position information (TSPI) data on the aircraft during flight tests. This receiver was replaced by the Collins 3A set in 1990.

The Central Inertial Guidance Test Division (CIGTF) at Holloman Air Force Base, New Mexico, became the focal point for managing DGPS reference systems in 1991, and Eglin Air Force Base in Florida became the third site to use DGPS for TSPI soon thereafter (a fourth site is mobile). Range accuracy of these systems is approximately 3–4 m, 3 drms, P-code. It is anticipated that high accuracy and dynamic flight tests will require P-code DGPS (C/A DGPS is 6–8 m, 3 drms). Velocity accuracy is better than 0.3 m/s.

8.5.1   Global Positioning System Translators

For instrumentation on a low dynamics vehicle, a single GPS is often sufficient. For a high dynamics platform the GPS is usually aided with an INS. When there is not room on the vehicle, a GPS receiver at a surveyed ground station is used with a *translator-equipped* test article (such as a missile). The missile receives L-band satellite GPS signals, translates them into the normal S-band, and then sends the data to the ground station, which also receives composite data from all GPS satellites. The overall system is now capable of DGPS accuracy after some data processing. Translators are cheaper than GPS receivers and provide a TSPI solution faster, but they require a high bandwidth.

8.5.2   High Dynamic Instrumentation Set

High Dynamic Instrumentation Set (HDIS) is a five-channel GPS set that fits in an AIM-9 missile pod and is used to collect data during flight tests. It can operate in the differential mode if required. When palletized, it is often referred to as the advanced range data system. Problems of vibration due to the sensor location are significant depending on the operating environment. Ground stations connected to a master facility for central control collect the data.

Test ranges have traditionally used Inter Range Instrumentation Group (IRIG) time standards that are broadcast throughout the range for time tagging of test events. GPS time is not IRIG time (see the next section) but is becoming increasingly popular because of its availability as GPS systems proliferate on the ranges and because it is continuous.

Kinematic CDGPS has not yet made a strong impact on test ranges, but it will in the near future. DGPS is being implemented at selected locations and used to generate TSPI data. The concept of a WADGPS national test range clearly has cost impacts worth considering, with an authorized SA P-code capability for selected military sites.

8.5.3 Specialized and Innovative Test Use of Differential Global Positioning System

Throughout the nation's test ranges a wide variety of ingenious applications are occurring using DGPS. NASA Dryden Flight Research Center used CDGPS on a formation flight of the SR-71 Blackbird and a F-16 aircraft to map the location of shock waves coming off the SR-71. In a similar way, CDGPS was used to record the dynamic interactions of an F-106 aircraft being towed behind a C-141 to simulate a towed space-launch of a rocket.

A capability called the *mobile pseudolite concept* exists at many military test ranges. The idea is for the test vehicle to carry a pseudolite and transmit signals to surveyed GPS units on the test range that are shielded from jamming and other interference. In this *reverse positioning concept*, multiple receivers take signals from the moving pseudolite to determine the test vehicle's position.

The U.S. Army uses a Deployable Force-on-Force Instrumented Range System (DFIRST) that results in highly realistic armored battle simulations at relatively arbitrary locations. One of its components is a 1-m jig with a GPS antenna at each end that clamps to the main gun barrel of a vehicle and records pointing accuracy within 0.2 deg. The complete system uses GPS satellites to position up to 78 vehicles executing shooter-target pairings and to keep track of who wins and who loses.

8.5.4 Inter Range Instrumentation Group Time

Computer-based data acquisition and signal processing systems have evolved from computers developed for more generic use and have real time clocks (RTCs) that are relatively inaccurate. They also are difficult to synchronize. Test ranges have traditionally used IRIG time, first mentioned in Sec. 2.1.3.

IRIG (G) time codes are 100-kHz codes for high-speed data acquisition to be broadcast and used throughout a test range for time monitoring or tagging test events. When multiple systems are all locked to the same standard, they share identical time codes, and records created by one system can be time correlated with records created by another system. Standard time broadcasts from Fort Collins, Colorado, are good choices in areas of strong reception where low accuracy is acceptable. Range telemetry stations have a central timing facility that receives UTC time and broadcasts it to telemetry equipped users. There are many IRIG codes but only three—IRIG A, B, and G—are recommended for use in high-speed data/time acquisition systems.

8.5.4.1 Real Time. IRIG time can be recorded on analog magnetic tape recorders along with analog test data by being amplitude modulated with 1000 cycles per frame (IRIG B = 1 kHz, IRIG A = 10 kHz, IRIG G = 100 kHz). The modulated time code as it is originally output from a time code generator is often termed "real time."

8.5.4.2 Linking global positioning system and interrange instrumentation group time. The UTC receivers and IRIG time code link data acquisition stations in clusters, simultaneously acquiring data from the same test. A missile test range, for example, may have launch, midrange, and impact clusters. Each cluster has GPS Receiver UTC time at 1 μs accuracy and a time code translator that decodes it. This time is put into the start or end of a data record. IRIG B time code can be time tagged to an absolute time within 1 ms. IRIG A and G time code can in certain implementations deliver 1 μs absolute accuracy.

8.6 Closure

GPS is possible because of the existence of technologies that promote and enable pseudoranging, precise modeling of the Earth and its gravity field, and incredible clock accuracy in the nanosecond range. Of these technologies, the availability of practical precise worldwide timing is proving to have a broad and lasting impact in its own right. GPS has made the world ask once again the timeless question: What is time?

Isaac Newton and his contemporaries viewed the heavens in motion as an absolute time standard, but it is now accepted that no true time measure exists in nature. The clock masterpieces of English craftsman John Harrison in the 17th century were the first to challenge the accuracy of the heavens, but the challenge was not taken seriously until the invention of the quartz crystal oscillator in the 1920s.

Since 1967 time has been defined as the duration of 9,192,631,770 cycles of radiation between two states of the cesium-133 atom. Clock accuracy in the modern era doubles approximately every two years, just like computer chip density. Time now defines distance—a meter is how far light travels in a specified fraction of a second (1 over 299,792,458 s)—and how accurate the heavens are in their cyclic motion (the Earth speeds up and slows down a few milliseconds each day).

Official time is international. It is averaged at the Bureau of Measures in Paris from over 50 world laboratories and, ironically, takes many weeks to be generated. GPS, via the Master Control Station in Colorado, is now the world's principle supplier of accurate, nearly instantaneous, relativity-corrected, real time.

9
Flight Testing Navigation Systems

A flight test is serious business. Avionics flight testing, of which navigation and guidance are a part, must be closely coupled with analysis. Often this requires accompanying laboratory simulations and/or tests that will themselves be validated from the flight-test data (some navigation simulations are capable of actually being driven by the flight-test data). This is not just an academic exercise, for there are an enormous number of variables in flight test (which is in itself very costly) that simply cannot be analyzed from only a few flights.

Avionics flight test is fundamentally technical risk management, and so the purpose of the tests must be clearly understood so that the resulting test data can be mapped into metrics useful in technical decision making. The error budget concept has been useful for decades in highlighting needed information and correlating it with data from other tests to provide technical risk reduction during systems development and test.

The objectives for this chapter are for you to 1) understand the practical considerations involved in testing a modern inertial navigation system and know how system testing is accomplished; 2) survey current test programs; and 3) describe the planned GPS integration cycle for DoD platforms.

9.1 Testing Autonomous Inertial Systems

The testing of aided inertial guidance systems was introduced in Sec. 5.4.3. There it was stated that a flight test of a system as complicated as an inertial guidance system should be measured against the following standard: Is it possible for the flight-test recorded data to be used to duplicate the flight test results on the ground?

This challenging standard is practically never achieved in practice because of time and cost constraints. But at the very least it should be determined why the test results will not be duplicable? The answers will help to illuminate the real purpose of the flight test. They also will highlight the instrumentation packages that are missing and identify data sets that are *vital* (as opposed to merely important). Finally, the answers will determine whether the test is being conducted to obtain a statistic such as CEP or to produce data for engineering analysis and development, such as the covariances in an error budget.

9.2 Inertial Navigation System Flight Tests

9.2.1 Central Inertial Guidance Test Facility

The CIGTF is part of the 6585th Test Group at Holloman Air Force Base, located approximately 90 miles north of El Paso, Texas, and 27 miles west of Alamogordo, New Mexico, near the White Sands Missile Range (WSMR). The area is seismically stable to the micro-g region, as required for precision guidance and navigation system testing.

The CIGTF complex consists of 10 buildings on approximately 227,000 ft^2 of real estate. The laboratory operates three centrifuge test beds, with and without counter-rotating platforms, subjecting test items to sustained acceleration environments up to 30 g, 50 g, and 100 g. CIGTF uses a variety of reference systems to provide accurate time space position information (TSPI) in conducting its laboratory, van, sled, and aircraft tests. CIGTF resources include a simulation laboratory, GPS satellite reference station, data analysis stations, and three state-of-the-art portable field jamming systems (PFJS). CIGTF uses the highly accurate sled track for testing precision guidance systems and validating post-mission filters and reference systems.

9.2.2 Two Phases of Inertial Navigation System Flight Testing

Testing of new INS systems within the Air Force is handled in two test phases. First, verification flight testing is accomplished by CIGTF in specially instrumented test bed aircraft. The second test phase is to evaluate the INS in its operational aircraft for a specific mission. Testing of this nature is usually performed by the Combined Test Force (CTF). Operational suitability and accuracy are the two areas that are specifically checked by the CTF. This chapter provides details of the testing performed by the CTF and then gives an overview of verification testing.

Inertial navigation system testing is often accomplished in conjunction with other aircraft tests. The checklist in Table 9.1 is intended to be a useful aid for conducting these tests.

9.2.3 Test Planning

9.2.3.1 Requirements definition. The INS test planning process begins with the definition of test requirements. Often the test requirements are determined by the *reporting* requirements. The responsible test organization (RTO) will establish the number of sorties, number of flying hours, and number of ground tests required, as well as the instrumentation and ground support requirements. Sufficient ground testing should be accomplished to establish the relationship of the ground accuracy to the in-flight accuracy and to validate the ground maintenance procedures.

For in-flight tests, plan on at least eight sorties of INS error data for every alignment method and navigation mode. Some flights, or portions of flights, should be dedicated to straight-and-level, nonmaneuvering profiles, and some should be dedicated to operational profiles. The number of sorties should be chosen so that meaningful statistics can be obtained during data reduction and analysis.

9.2.3.2 Test method. The INS test method must be based on sound theoretical considerations. Because modern inertial navigation systems can take a variety of forms, a test director or test planner must know the system: gimbal or strapdown platforms, mechanical or laser gyros, aided or unaided modes. Each of these system implementations has its particular characteristics and problem areas.

An INS considered for flight testing will have a considerable amount of test data already available. Test data may be from the INS manufacturer's development

Table 9.1 INS program manager's checklist

Verification (accuracy) tests:

1. Ensure tracking facilities or airborne reference systems are available to provide TSPI to the accuracy required for determining spec compliance. Facilities available include cinetheodolites, radar, and laser tracking. Airborne reference systems include astro-inertial systems, CIRIS, global positioning system, photography, and air combat maneuvering (ACMI) ranges. The system or facility selected will be dependent upon budget, time constraints, location, and accuracy requirements.
2. Ensure that flight profiles are designed to identify all error sources.
3. Are sufficient flights dedicated for INS testing to ensure that the required statistical confidence levels can be met?

Suitability tests:

1. Do the flight profiles cover the entire operational envelope?
    a) *g* limits–both positive and negative
    b) Airspeed/Mach
    c) Maneuvers
    d) Environmental-temperature, size, electrical power, weight
    e) Time constraints
        1) alignment time vs scramble requirements
        2) length of flight vs allowable errors
2a. Are the intended operational uses sufficiently tested?
    a) DISPLAYS
    b) INPUT TASK AND LOCATION
    c) READABILITY AND LOCATION
2b. Does the system properly integrate with the other systems?
    a) AUTOPILOT
    b) BOMB NAV
    c) NAV COMPUTER
3. Is the mean time between failures (MTBF) acceptable and operationally arrived at?
4. Is the mean time between repair (MTBR) acceptable and operationally arrived at?
The following considerations should be addressed during the test:
    a) Who will perform the maintenance?
    b) Are the technical orders provided sufficient both for the maintenance and operational people?
    c) Are all types of alignments being tested?
    d) Are all degraded alignments that represent operational requirements being tested?
5. All alignments should be made on different headings to ensure a good check on the alignment feature of the system.
6. Is the alignment position-on the ramp free from magnetic disturbance?
7. Sufficient time between runs should be allowed for system cool down (usually 12 h).
8. Ground runs should be made prior to and after a flight using one alignment. This allows identification of static errors.
9. Data should be taken every 5 min during the flight. This data must be taken from the INS as well as the tracking facility. This allows a very accurate plot of system errors.

(*Cont.*)

Table 9.1 (Continued)

10. All legs of the flight should either be 84.4 min of fractions of this time, i.e., 21.1, 42.2, etc.
11. Are there any differences between operating on ground power and aircraft power?
12. Does the operational environment require a time delay on power interruptions?
13. Are operator errors minimized?
14. Have environmental aspects been addressed during the program? This is especially true if the aircraft with its INS must operate in either arctic or other harsh environments.

testing, from testing conducted by the CIGTF, from the avionics development and integration test program of the airframe contractor, or from test reports on other aircraft with the same inertial system installed. Any system test should include consideration of data available from earlier tests. These data sources can be extremely helpful to determine specific test objectives and methods.

A key element in the proper design of any test is to understand the characteristics of the error. A conventional, unaided local vertical inertial navigation system exhibits bounded and unbounded errors that vary with the Schuler oscillation. Aiding is often employed in an attempt to bound the error. Knowledge of the error budget for a given system is of great importance to anyone planning a test because it helps focus on particular system characteristics. In general, you should test for the effects of temperature, accelerations, maneuvering (including acrobatic) flight, and alignment heading on position and velocity error. In some systems, moving from one hemisphere to another may present a problem, either in alignment or navigation, or both.

A major consideration in testing an inertial navigation system is how well it interfaces with the many other systems in the aircraft that require INS inputs. For example, sensors require attitude stabilization, ground speed, and true ground track. Additionally, the displays used to present INS data are usually separate subsystems, often designed and built by a company other than the INS manufacturer. Individual subsystems can be "hard wired" together with discrete connections for each parameter, or they can be tied together over a multiplex bus. In either case, the human factors assessment takes on increased importance when multiple subsystems are integrated. Data displayed in the cockpit must be verified by an external source.

The strapdown inertial platform is becoming the system of choice in today's navigation systems. Although a strapdown platform has the same characteristic gyro and accelerometer errors as a gimbaled platform, these errors act along the aircraft body axis only. As a result, the clear correlation between the error source and the position, velocity, and azimuth errors is reduced. Additionally, a strapdown system can be expected to have a larger share of its error budget allocated to computational errors than a gimbaled system.

Avionics systems have evolved to the point where the inertial measurement unit (IMU) is but one component of an integrated navigation system. This approach allows the use of less precise (and less expensive) inertial platforms, which, when combined in a Kalman state estimator with other navigation sources such as GPS, Doppler, air data, etc., can yield highly accurate navigation information. Error sources such as gyro and accelerometer bias and scale factor errors are often modeled internally in the KF, thus they can be estimated during flight and increase the accuracy of the integrated system.

An important consideration in data reduction is the accuracy of the tracking data. Flight-test ranges have tracking radars for measuring and recording aircraft actual position, but they may not be sufficiently accurate. In this case, additional methods of determining aircraft position, such as GPS, dedicated transponders, photography, and laser tracking should be investigated. Remember that the source of navigation truth data may be questioned.

For a cruise flight test PVAT data should be taken every 5 (not to exceed 10) min and much more frequently during maneuvers. An automatic data recording system is highly desirable for INS testing. Besides relieving the operator of the burden of recording parameter values at frequent intervals, an automatic system offers distinct advantages: 1) the possibility of automated data reduction and 2) the possibility of recording additional parameters besides position and velocity. For automatic systems ensure that the recording system is capable of storing data for the duration of the flight. In addition, some method of correlating the onboard data with the truth data is essential. This generally takes the form of continuously recorded IRIG time and/or audio tones that serve as event markers.

Tests require some form of data card, both to record data and to aid in cataloging and organizing results. The data reduction scheme should be known and used in the preparation of the flight data card. Essential items should be highlighted.

9.2.4 Test Conduct

9.2.4.1 Ground tests. INS ground testing generally has two parts: alignment tests and baseline position accuracy tests. For alignment tests, the objective is to exercise all available alignment modes. Results from a number of like tests can be combined to determine the mean time to align for a given mode. Record the timing of align phases under different conditions of initial aircraft heading and ambient temperature. Allow the system to thoroughly cool (at least 12 h) between alignment tests. After shutdown, move the aircraft to a new heading to fully exercise the self-alignment capability. Document local wind conditions and any buffet or movement of the aircraft during alignment. Also note whether the aircraft is inside or outside of the hangar and clear of large metal objects such as ground power units that could affect the alignment. Finally, assess the suitability of the system for any special mission requirements that may exist (i.e., helicopter operations, scramble, etc.).

Ground navigation tests should consist of at least eight runs of 42 min in NAVIGATE mode while stationary. Accomplish ground testing on both internal and external power, if possible. Park the aircraft over surveyed coordinates and record system position and velocity every 5 min. In addition, post-flight ground runs should be accomplished in conjunction with in-flight tests to identify system errors. Plan on at least a 21-min post-flight run time for the ground tests, using the same alignment as the in-flight tests (i.e., do not shut down after flight). Reduce the data in the same way as in-flight data are reduced. Finally, evaluate the built-in-test/system integrated test (BIT/SIT) features of the system, with emphasis on external verification of internal fault analysis.

9.2.4.2 In-flight tests.
In-flight tests should consist of both nonmaneuvering and operationally representative profiles. For nonmaneuvering tests attempt to limit excitation of errors to one axis only by flying East/West and North/South runs on separate missions. This will help establish a baseline error profile that can be compared with ground data as well as operationally representative mission profiles. Fly at least eight test flights.

One of the most important elements in the testing of integrated systems is the qualitative assessment of operational suitability. This evaluation is primarily a human factors evaluation. Assess the operator/system interface, including displays and controls, data entry and extraction, and location and functional grouping of components. Qualitatively evaluate the INS interface with other systems, such as head-up display (HUD), weapons computer, and autopilot. Establish some method of quantifying your qualitative data by using questionnaires, workload analyses, or Cooper–Harper ratings.

The Cooper–Harper ratings are well known in flight test. For a specific mission task and criteria, a Level 1 evaluation (assigned rating between 1 and 3) implies that the system under test is "satisfactory." A Level 2 evaluation (assigned rating between 4 and 6) implies the system is "acceptable" but not "satisfactory" in some way. A Level 3 evaluation (assigned rating between 7 and 9) indicates that for the mission task as defined the system is neither "satisfactory" nor "acceptable." The numerical rating within each Level is often used with operator comments to describe more completely system operation and operator workload during test. These numerical ratings are often averaged, but care must be exercised when this is done across Level boundaries. *User comments are essential* in understanding rating variations between different Levels.

9.2.4.3 Maintaining control of the test.
The integrity of your test must be maintained against outside influences. A legitimate test must be independent and free of the influence of those with preconceived opinions about the test article. Guard against restrictions placed on normally allowable maneuvers, flight conditions, and cool-down time. If INS tests are conducted too early in a program, i.e., during envelope expansion, the flight profiles may not be operationally representative. Ensure that enough flights are flown for the data to be statistically valid.

Quick-response repair capability is a must for system tests, particularly new or prototype systems that can have a relatively high failure rate. Midstream maintenance, however, often leads to problems of configuration control that can invalidate much of the data. This is particularly true in a developmental test where changes can occur rapidly. Establish a rigorous procedure for configuration control early in the program, and stick with it. Initiate a comprehensive cataloging procedure for tracking data and reports, and avoid the tendency to collect vast amounts of unidentifiable data.

INS testing is usually done in conjunction with other tests to maximize sortie productivity. This requirement can have an impact on aircraft availability, alignment procedures, and in-flight profiles. Do not allow an advocate of the system to reduce the data. Be in charge of data reduction and be prepared to explain the data reduction process.

Repeatability is fundamental to any test. The way to ensure repeatability is to thoroughly document the test procedures and be disciplined in the taking of data.

Follow configuration control procedures. If the test extends over a long period of time, the chances of losing control of configuration and losing data integrity increase.

9.2.5 Data Reduction

9.2.5.1 Geometric mean/root mean square method. The data from the flight tests yield radial error as a function of time. This error is found by comparing measured INS position with actual truth position (obtained from other sets of independent measurements). The result is a set of eight or more curves (one for each flight) of radial error vs time, as shown in Fig. 9.1.

A common method to reduce two-dimensional position data is the geometric mean/root mean square (GM/RMS) method. If there are n samples of data, each with its own value for radial error r_i, then the *geometric mean* is defined to be

$$\text{GM} = \left(\prod_{i=1}^{n} r_i\right)^{1/n} \tag{9.1}$$

and the *root mean square* (rms) is given by

$$\text{rms} = \sqrt{\frac{1}{n}\sum_{i=1}^{n} r_i^2} \tag{9.2}$$

then the ratio (GM/RMS) is given by

$$\gamma = \frac{\left(\prod_{i=1}^{n} r_i\right)^{1/n}}{\sqrt{\frac{1}{n}\sum_{i=1}^{n} r_i^2}} \tag{9.3}$$

This ratio, which must be less than or equal to one, is used to evaluate system performance in terms of percentile error predictions. For example, if it is desired

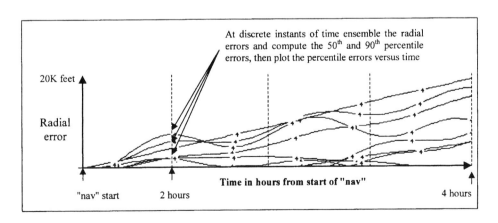

Fig. 9.1 Typical radial errors for an ensemble of flights.

to be 90% confident that radial error is less than a certain specified value R(90%), then that value can be found from a statistical formula. These formulas have been tabulated for various confidence levels in Fig. 9.2. Returning to the example, if the GM/RMS ratio is 0.6, then from Fig. 9.2

$$\frac{R(90\%)}{\text{rms}} = \gamma = 1.62 \tag{9.4}$$

or equivalently, there is a 90% probability that the position error will be less than

$$R(90\%) = \gamma \cdot (\text{RMS}) = 1.62 \sqrt{\frac{1}{n} \sum_{i=1}^{n} r_i^2} \tag{9.5}$$

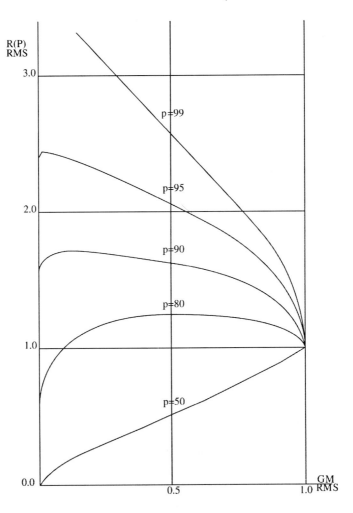

Fig. 9.2 Percentile error curves.

Calculate the GM/RMS value for the ensemble at discrete times. Determine R(p)/RMS from Fig. 9.2 for whatever percentile is desired. This value R(p)/RMS is the "normalized pth-percentile level of error." Commonly, the 50th or 90th percentile level is specified. This normalized value may be converted by multiplying by the RMS value as in Eq. (9.5).

The result, R(90%) or the "90th percentile level of error," means that the position error in a given flight will be less than this value on 90% of flights (if the number of flights is large). It is also correct to say there is a 90% probability that the position error in a given flight will be less than this value.

Compute the 90th percentile level of error, R(90), for discrete times throughout the mission. Plot R(90%) vs time and fare a curve through the individual points. This represents the accuracy of the system to a 90% confidence level. Most INS contractual guarantees are expressed in terms of R(50%). For determining specification compliance, compute and plot the R(50%) value for the same data set. Plot the specified accuracy on the same plot for comparison.

9.2.5.2 Circular error probable. The direct CEP plot of normalized terminal error is a popular method of data presentation. Radial error at a particular time or at the end of a flight is determined by comparing INS position with known coordinates of the aircraft parking spot. This value of radial error is normalized at some value in time, usually 1 h of mission time. Limits must be placed on what constitutes a suitable sortie duration for this calculation. Sortie durations that vary from the desired duration by more than a specified percentage should not be accepted. This percentage limit must be specified in the test planning phase if this method of data presentation is planned. A large number of data points are required for this method to yield meaningful results.

A sample terminal error CEP plot is shown in Fig. 9.3. A circle with radius equal to the specification CEP is overlaid on the plot of normalized latitude and longitude errors. Analysis of this data is best done by computer. The computer forms the ratio of points within the circle to total points. After the initial calculation the computer should examine the distribution of error points with respect to the value of the calculated CEP. If the terminal error from any of the flights exceeds a three-sigma value of the CEP, that data point is considered an "outlier," and the CEP calculation is repeated.

9.2.6 Verification Testing

Verification testing of any new INS is accomplished by the 6585th Test Group at Holloman Air Force Base, New Mexico. The primary purpose of verification testing is to provide a basis for comparative analysis of the performance and operational suitability characteristics of a series of theoretically similar navigation systems. This is done by obtaining comparative results under nearly identical test conditions in a minimum amount of time. At the same time reliability testing is being accomplished to verify that system or component reliability is equal to or better than a specified minimum value. The basic verification test outline from this guide is shown in Table 9.2. The objective of the verification is to establish a level of statistical confidence in the performance of the system for its primary operating mode in a typical operational environment.

174   INTEGRATED NAVIGATION AND GUIDANCE SYSTEMS

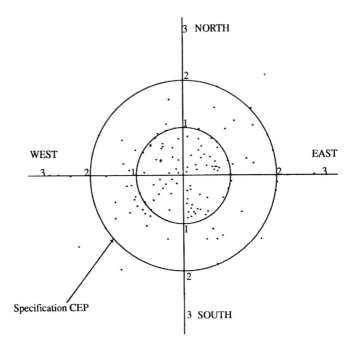

Fig. 9.3   Typical terminal error CEP plot.

9.2.6.1   Verification test flight profiles.   The perfect verification flight profile would accomplish the following: 1) provide a constant readout of the acceleration, velocity, and position errors; 2) identify the various error sources; 3) ensure that no errors were masked or lost during the conduct of the test due to flight paths; and 4) allow all competing systems to fly over it to provide a comparison.

In reality, flight constraints must be considered. These include the inability to measure acceleration without biasing it, the difficulty of determining the source of an error, the limitations in accurately determining the position of the vehicle by an independent source, and the fuel and maneuvering limitations of various aircraft.

When these constraints are considered, a practical set of flight profiles includes the following characteristics:

1) The accuracies of radar for tracking the aircraft are limited to approximately 200 ft at 30,000 ft altitude, and data turnaround is 2 to 3 days. Surveyed checkpoints that are photographed from the air vary in accuracy with altitude, and data are available within 24 h.

2) The current reference system for flight profile control at Holloman Air Force Base is the Completely Integrated Reference Instrumentation System (CIRIS). CIRIS provides a highly accurate position, velocity, and altitude reference over a long flight path for real-time use. CIRIS is carried as an instrumentation pallet on transport test aircraft or center-line pod on fighter test beds. Reference data from CIRIS is accurate to 13 ft in the horizontal axes for position and to 0.1 ft/s for velocity.

Table 9.2 Basic verification test outline

Phase	Testing	Totals
Phase I-A	Standard ground tests	
	Functional checkout	
	Static NAV tests	
	Scorseby tests	
	Heading sensitivity tests	
Phase I-B	Special ground tests	7–15 tests/
	Special analysis tests	4–8 weeks
	Special application tests	
Phase II	Transport flight tests	17–22 flights/
	Functional checkout flights	54–72 hours/
	West 84-min cruise flight	8–12 weeks
	North 84-min cruise flight	
	East 84-min cruise flight	
	Terminate at distant point	
Phase III-A	Functional checkout flights	19–24 flights/
Helicopter flight tests	6 East 42-min cruise & return	29–39 hours/
	6 North 42-min cruise & return	8–10 weeks
	6 East 42-min terrain mapping mission & return	
	Up to 6 special analysis flights	
Phase III-B	Functional checkout flights	21–24 flights/
Fighter Flight tests	6 West 42-min cruise & return	36–41 hours/
	6 West 42-min cruise/simulate ordnance delivery & return	8–12 weeks
	6 West 42-min cruise/simulate air combat & return	
	2 to 4 special analysis flights	
Phase III-C	Functional checkout flights	7–13 flights/
Cargo flight tests	2 or 3 West 168-min cruise & return	48–84 hours/
	2 or 3 East 168-min cruise & return	5–9 weeks
	1 East terminate at distant point	
	Up to 3 special analysis flights	
	2 or 3 North 168-min cruise profile & return	

3) Because of the Schuler cycle, 84.4 min legs should be flown; however, some fighter and helicopter aircraft cannot fly that long. For these aircraft a profile should have multiples of 21.1 min legs.

4) Because error analysis is a very complicated process, the profile must help isolate the various error sources. This is accomplished by exciting only one channel of the platform at a time and by conducting ground runs to determine the static errors. The verification flight profiles thus may look as depicted in Fig. 9.4, with either 84- or 168-min outbound legs. Routes start with a ground alignment and

Fig. 9.4 INS verification flight profiles.

end with a 21-min ground run after flight. Fly cardinal directions for 84 min when practical and overfly the same path to see if errors cancel.

9.3 Testing Integrated Navigation Systems

9.3.1 Test Program Survey

9.3.1.1 Antijam global positioning systems technology flight test. GPS is creating a new era of weapon guidance. GPS will provide guided weapons with an adverse weather, worldwide, day or night, high accuracy navigation and guidance capability. However, a GPS receiver alone cannot provide attitude information at an adequate rate for stable weapon control, and hence, an IMU is required.

Tightly coupling the output of the GPS and IMU (see Sec. 7.1.3) results in an optimal navigation system. One of the primary concerns of using GPS on weapons is the threat of enemy electronic counter measures (ECM) attempting to jam the radio frequency (RF) signals transmitted by GPS satellites. The intensity of jamming encountered by a weapon is more severe than that of any aircraft GPS system because the weapon flies in closer to the jammers, which may be collocated near the target. The Antijam GPS Technology Flight Test (AGTFT) program is structured to demonstrate and jam GPS/INS weapon guidance in a jammed environment.

The military has a long history of developing GPS and IMU guidance technology for weapons. The first weapon GPS receiver was developed in the late 1970s. More recently, the Tactical GPS Antijam Technology (TGAT) program, completed in 1992, developed an adaptive GPS filter/antenna that simultaneously accomplished jammer discrimination in both the spacial and temporal domains. Spacial discrimination was accomplished by phasing and weighing signal and jammer energy received from a small number of "patch" antenna elements to produce the proper antenna null direction. Temporal discrimination was accomplished by phasing and weighing delayed segments of the composite GPS wave received by the antenna. Jam resistance realized with TGAT hardware was impressive against both broadband and narrow-band jammers.

More recently the military has developed the GPS Guidance (THAAG) program to advance the spacial and temporal antijam techniques. This program integrates antijam techniques with a GPS receiver, tightly coupled with a weapon grade IMU, into a single antijam (AJ) GPS/INS weapon guidance set, again demonstrating significant jam resistance.

The objective of the antijam GPS flight-test program is to identify and test a low-cost high antijam GPS/INS weapon guidance system for joint direct attack munition (JDAM) type weapons. Flight testing is scheduled to begin in 1998.

9.3.1.2 Tactical miniature inertial measurement unit. The Weapon Flight Mechanics Division (WL/MNA) at Eglin Air Force Base, Florida, is investigating novel means of controlling the flight of air-to-air missiles to dramatically increase missile off-boresight launch capability and high angle-of-attack flight. They also are involved in a joint program with the Army to develop a low-cost antiarmor submunition that can discriminate between truck and tank targets in both clear and adverse weather and against various countermeasures. Future air-to-air missile guidance systems will benefit from their development of tactical grade miniature IMUs.

These advanced IMUs will be more accurate, more reliable, and significantly less expensive than the units they will replace. Complementary development is being conducted in a tactical weapon high antijam GPS guidance program to couple a low cost antijam antenna/filter system with a low cost tactical grade IMU. This technology will allow future air-to-surface weapons to receive GPS navigation signals to within less than one mile of targets protected by collocated jammers, thereby insuring the highest possible kill probability.

9.3.1.3 Micromechanical inertial measurement unit. Also being developed at Eglin Air Force Base is a very small and inexpensive three-axis micromechanical gyro system. The military is funding a contractual effort with The Charles Stark Draper Laboratory to accomplish this work. Under the current contract, Draper Laboratory has fabricated and tested the first three-axis microelectro mechanical (MEM) system ever assembled utilizing three MEM gyros. Each of the hybrid MEM gyros is on a single printed circuit board that is fixed to a conservative 5-in. cube layout for maximum access and testability of the prototype system.

9.3.1.4 Advanced tactical inertial measurement unit. The objective of the advanced tactical IMU (ATIMU) effort is to develop a miniature (15 in.^{3}), extremely low-cost (less than $2000) tactical missile grade IMU using innovative MEMs technology that will easily lend itself to low cost manufacturing processes. This effort should result in a highly reliable "next generation" IMU with a gyroscope drift rate performance requirement of 1 deg/h and an accelerometer bias stability of 300 μg.

The ATIMU should also have a substantially lower unit production cost than that of ring laser gyro (RLG) and fiber optic gyro (FOG) based IMUs. The goal is to integrate the MEMs and the required electronics into a three-axis IMU package that will exhibit tactical missile grade performance. The ATIMU will be suitable for use in a wide variety of tactical munition systems.

9.3.1.5 Fiber optic gyro inertial measurement unit. The FOG IMU effort is a small (25 in.^{3}), low-cost (less than $6K) tactical missile grade IMU using interferometric FOG technology and integrated silicon accelerometers (ISA). The technical approach is to develop the FOG IMU utilizing previously developed FOG and ISA technology. The FOGs will be of the interferometric type consisting of a low-cost light source, detector, fiber coil, I/O chip, couplers, and other supporting electronics. A breadboard will be designed, fabricated, tested, and refined. These refinements will be incorporated into a miniature brassboard IMU.

Until recently, fiber-optic sensors had to be much larger than the optical source wavelength because of the capacitive readouts. However, a new type of optical sensor, proposed by Brent Little at the Massachusetts Institute of Technology, shows theoretical displacements resolutions near 0.01 pm. This technology, if fruitful, will open the door to MEMs with readout limits below 0.01 deg/h.

9.3.2 Global Positioning System Integration Cycle

GPS is the largest avionics procurement and installation program in the history of DoD. When completed, GPS will have been integrated and installed in over 18,000 aircraft (over 100 different type/model/series) plus a majority of other weapon systems (435 ships, 35,000 vehicles, numerous weapons). Integrating GPS into aircraft systems, whether as a modification to an existing integration scheme or as a part of a new system, is a technically challenging, time consuming, and costly effort.

GPS is being integrated into over 40 different U.S. Air Force aircraft. Most integration programs are currently in the design and test phase. Integrations are complete for the EF/F-111, MC-130, AC-130, MH-60, MH-53, U-2, and RC-135. Installations are ongoing on the B-52G, EC-130, and C-130, aircraft retrofit programs. In addition, the F-16 C/D, C-17, E-8, B-2, C-130, and MH-60G in-line production aircraft are receiving GPS on the production line. To date, approximately 16% of the U.S. Air Force fleet is equipped with GPS. By the year 2000, 91% of the fleet will have GPS. Recent increases in force structure and realignment of the T-38 trainer aircraft GPS modification with another avionics upgrade account for the remaining 9%.

9.4 Closure

When Lt. Jimmy Doolittle began his blind-flying experiments in the late 1920s that were sponsored by Harry Guggenheim, Professor William Brown took leave from his teaching duties at the Massachusetts Institute of Technology (MIT) to work with Elmer Sperry on flight instruments for Lt. Doolittle's aircraft. The Aeronautics Department had difficulty finding a substitute for Professor Brown who could teach the new course "Aircraft Instruments" and as a last resort turned to a young research associate who had a degree in psychology from Stanford and an electrochemical engineering degree from MIT. He had taken more courses for credit at MIT than anyone else to date (and possibly since), and so he was assigned the class and given a box of slides plus a bushel basket full of Professor Brown's instruments.

The research associate was Charles Stark Draper, who for the next four decades would revolutionize the field of guidance, navigation, and control. "Doc" Draper, a pilot in his own right, improved the design for Sperry's turn indicator. But he could not market his design, and so he spent World War II designing gyroscopic sights for weapons systems. This culminated in an inertial bombing system, which in turn led to the first demonstrated inertial navigation system, SPIRE, that flew from Boston to Los Angeles in the late 1940s. From this success came inertial navigators for fighter aircraft, submarines, and of course for the Apollo vehicles that enabled Neil Armstrong to walk on the moon. The Draper Laboratories continue to serve NASA with avionics systems for the Space Shuttle.

I met "Doc" Draper in 1983, a few years before he died in 1987, and was impressed by his continued vitality and curiosity. He showered me with instruments for an educational laboratory that I was starting and emphasized that engineering education should never be solely theoretical. He had always involved his students, many of whom became high-ranking officers in the military, in ongoing projects. He was a natural at flight testing complicated systems and was a master of hands-on education. MIT's motto "Mind and Hands" could not have found a more fertile and receptive intellect than that of Charles Stark Draper (October 2, 1901–July 25, 1987).

10
Navigation Computer Simulations

This chapter is accompanied by a PC computer file called AINS.EXE where AINS stands for *aided inertial navigation systems*. This file, when executed, will self-extract into a computer textbook by Stanley Schmidt and accompanying computer simulations that are designed to provide insight and discovery into the operation of inertial navigation systems and Kalman filters (KF). This chapter will describe in detail the steps to be followed to install the program and run the computer simulations. The exercises are designed to illustrate the necessity of computer simulations for the error modeling of integrated navigation systems and for the propagation of errors using the Kalman filter.

From the handout for this chapter you will be able to do the following:

1) Demonstrate inertial system error propagation for initial position, velocity, and attitude misalignment errors; discover the effects on system performance for a combination of these types of errors.

2) Demonstrate KF error propagation for a two-dimensional tracking example; develop and illustrate schemes to cause filter divergence; discover the weak and strong points of filtering algorithms and predict the impact on airborne navigation applications.

10.1 Installation of AINSBOOK on Your Personal Computer

The AINSBOOK software by Stanley Schmidt comes as a file called AINSBOOK.EXE that is a self-extracting installation of AINS plus a README file. Place these files in a directory of your choice, preferably the C:\AINS directory in your computer, which will be created automatically if defaults are followed. The program will install itself when it is run by double-clicking on it, either in the Windows or DOS environment, and will ask you where you want the program located. It is highly recommended that you look at the README file that summarizes the installation procedure and shows how to get help if there is a problem.

To run the complete AINS program, run the executable file called AINS.EXE from either the Windows or DOS modes (DOS is preferred in a Windows 3.0 environment, but Windows 95 and Windows NT run AINS.EXE well enough without going to the DOS prompt). If you have created an AINS icon on your desktop for the program, then right-clicking on it and selecting properties, screen, will allow you to determine whether you desire a full or partial screen for the program.

If you are successful, the screen in Fig. 10.1 will be shown on your screen; if not, get help! After you hit the spacebar (SB), followed by a carriage return (CR), you will be in the text mode of the narrative on inertial navigation and Kalman filtering written by Schmidt (who, by the way, wrote his own word processor for the screen display). Use the Page Down key to explore the leader, then hit the F1

AIDED INERTIAL NAVIGATION SYSTEMS
USING EXTENDED KALMAN FILTERS

A compilation of formulations, designs
and implementations of Kalman filtering
for use
as a reference for
practicing engineers and
computer scientists.

prepared by

Stanley F. Schmidt, Ph.D.
Consultant

Version 1.2
(c) Copyright 1990 - 1997
Hit space bar to continue

Fig. 10.1   AINS program.

key to see how the help is set up. Note that the simulation graphics mode may be entered directly by using the F2 or F3 keys, as desired, and that bookmarks may be entered by the user in the text.

Printing is not supported from the AINS program, but the Alt-PrintScreen keys, hit simultaneously, should copy the screen image in Windows for pasting into a word processing document. If this feature does not work properly, try hitting the Alt-Enter keys simultaneously before the Alt-PrintScreen keys. The Alt-Enter keys may also assist in obtaining the proper graphics display of certain symbols. In the Windows NT environment, it may not be possible to copy the screen image while maintaining the proper display for symbols. If the screen freezes in the NT environment, Alt-Enter should resume operation. In the NT environment, Control-PrintScreen is a better choice for screen saving than Alt-PrintScreen from the graphics mode of the AINS program.

Feel free to become familiar with the computer display of the contents of this remarkable program, but realize that the notation is different than that used in this text and that Schmidt writes technically in depth. He provides user assistance at the bottom of the screen and has set up the Esc key to get you out of most difficulties. Chapters VI and VII of Schmidt's text contain detailed explanations of the INS error modeling program and of the KF tracking simulation. Introductions to these topics follow.

10.2   AINSBOOK Inertial Navigation System Error Analysis Simulation

You may engage the graphics mode of the INS error simulation for a strapdown INS from anywhere in the Table of Contents by hitting the F3 key. As an alternate

NAVIGATION COMPUTER SIMULATIONS

approach, from the Table of Contents highlight Chapter 7.5 with the arrow keys (see Fig. 10.2) and CR, then continue hitting the PageDown key until the instructions "HIT PgDn for the tutorial" appear just before the computer goes into the interactive graphics mode. To return to the text mode, hit the Esc key and follow the screen instructions. Schmidt has set up the program so that hitting ESC will get you out of most predicaments. You are now ready to complete the laboratory exercise on the AINSBOOK INS error analysis simulation.

10.2.1 Mechanization Equations for Strapdown System Simulation

The Fundamental Equation of Navigation (FEN) was presented in Chapter 3 of this text, and it is written here in C_v coordinates (vehicle-carried vertical frame) as

$$\dot{v}_{k,v} = \frac{f_{\text{local},v}}{m} - [\Omega_{(v/i,v)} + \Omega_{(e/i,v)}]v_{k,v} \tag{10.1}$$

which can be computed in C_v coordinates using the block diagram shown in Fig. 10.3. The local force f_{local} equals the local gravity force plus the output of the specific force sensors (accelerometers). In a strapdown system the accelerometers are in C_b coordinates that must be transformed using C_b^v if FEN computation occurs in C_v coordinates. The INS groundspeed v_k is converted with Eq. (3.15) to latitude rate, longitude rate, and altitude rate. These rates are used to determine the computed rotation rate of the vehicle-carried vertical frame as the aircraft moves over Earth's surface, $\omega_{v/e}$, using Eq. (2.43). In Schmidt's AINSBOOK the FEN is implemented as in Fig. 10.3 using a different notation scheme.

The FEN mechanization may then be embedded in the computation for the rate of change of direction cosine matrix (DCM) $\dot{C}_v^b$, which when integrated provides the DCM C_v^b and its inverse C_b^v. This rate of change is found from Eq. (3.23) in Sec. 3.3.4 as

$$\begin{aligned}\dot{C}_v^b &= -\Omega_{(b/v,b)}C_v^b \\ &= -[\Omega_{(b/i,b)} - \Omega_{(v/i,b)}]C_v^b \\ &= -\{\Omega_{(b/i,b)} - [\Omega_{(v/e,b)} + \Omega_{(e/i,b)}]\}C_v^b \end{aligned} \tag{10.2}$$

that can be mechanized as shown in Fig. 10.4.

Note from the figures how gyro drift error, an unwanted but always present component of $\Omega_{b/i,b}$, directly affects the computation of the orientation of the strapdown axes, which will result in tilt errors exciting the FEN equation. Accelerometer errors also directly excite the FEN equation via $a_{\text{sf},b}$ and will show up in the navigation outputs. Strapdown system errors commonly referred to as coning or skulling are not covered in this development.

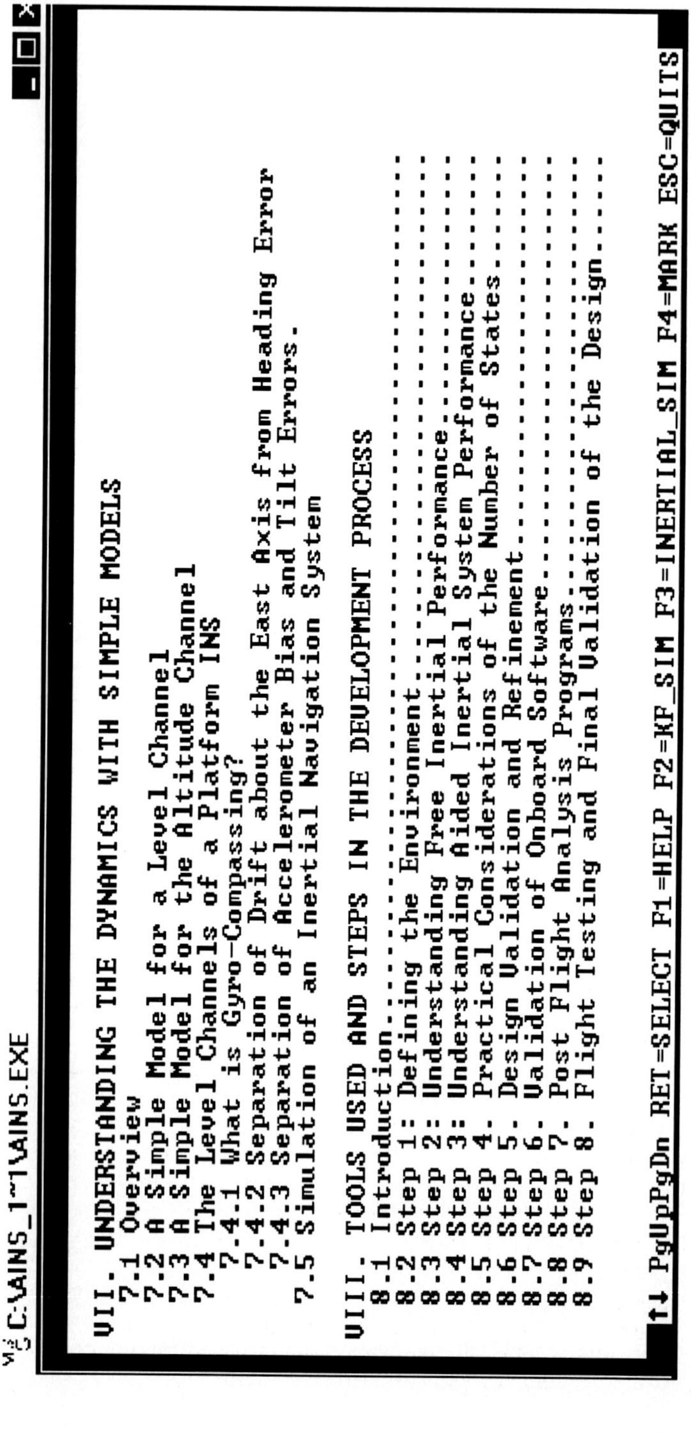
C:\AINS_1~1\AINS.EXE
VII. UNDERSTANDING THE DYNAMICS WITH SIMPLE MODELS
7.1 Overview
7.2 A Simple Model for a Level Channel
7.3 A Simple Model for the Altitude Channel
7.4 The Level Channels of a Platform INS
7.4.1 What is Gyro-Compassing?
7.4.2 Separation of Drift about the East Axis from Heading Error
7.4.3 Separation of Accelerometer Bias and Tilt Errors.
7.5 Simulation of an Inertial Navigation System
VIII. TOOLS USED AND STEPS IN THE DEVELOPMENT PROCESS
8.1 Introduction
8.2 Step 1: Defining the Environment
8.3 Step 2: Understanding Free Inertial Performance
8.4 Step 3: Understanding Aided Inertial System Performance
8.5 Step 4. Practical Considerations of the Number of States
8.6 Step 5. Design Validation and Refinement
8.7 Step 6. Validation of Onboard Software
8.8 Step 7. Post Flight Analysis Programs
8.9 Step 8. Flight Testing and Final Validation of the Design
↑↓ PgUpPgDn RET=SELECT F1=HELP F2=KF_SIM F3=INERTIAL_SIM F4=MARK ESC=QUITS

Fig. 10.2 AINS program Table of Contents.

NAVIGATION COMPUTER SIMULATIONS

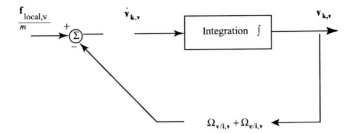

Fig. 10.3   FEN block diagram.

10.2.2 Generating Inertial Navigation System Error Plots for a Strapdown System

The equations and mechanization for a stationary Earth-fixed strapdown system are embedded in the interactive computer simulation. Recall that the simulation may be initiated from anywhere in the Table of Contents by hitting the F3 key. If you have entered the simulation without difficulty, you should obtain a screen that looks like Fig. 10.5 after selecting the F1 help key. Note the commands at the bottom of the screen and remember that the Esc key will exit the graphics mode for the text mode. Scrolling through the help window will explain in detail how to run the simulation and scale the plot outputs.

INS case studies are run using the F6 or F7 keys, and you should try the F6 key to investigate how the plot variable and units appear on the plot. The program is set up by default to plot East velocity error in units of feet per second that results from an initial East velocity error of 5 ft/s (see Fig. 10.6). The coordinates used by Schmidt are the 1 axis for the local vertical (up), the 2 axis for the East direction, and the 3 axis for North, making a right-hand set. Note from the figure that an initial East velocity error is equivalent to an initial tilt error rate and a position error rate about the North axis.

For the interactive INS error plot simulation, you are allowed to plot velocity errors, position errors, heading errors, and tilt errors by selecting the appropriate variable with the F2 key. (Scaling is not automatic and will be required for some selections.) Using the F5 key, you may set the values for a wide variety of errors as forcing functions, one at a time or in combination, and in addition input accelerometer and gyro biases and scale factor errors. The F2 and F5 keys are normally used to make these selections, and it is very important to realize that Page Down is activated when viewing these displays and input screens, so that more variables are available than the ones being viewed. The INS simulation is *deterministic*, and normally using the input F5 key you will select the deterministic error sources driving the strapdown INS. The F6 key runs the simulation with your selection.

The F7 key runs the simulation with your input settings as one-sigma rms input to a random number generator that, in turn, outputs a sample value for the simulation. In other words, running the default INS error simulation with the F7 key uses an East velocity error sample as input from a population with a one-sigma value of 5 ft/s. Thus you will get different error plots every time you hit the F7 key, just as you would in test runs with a random but fixed East velocity error driving the INS.

186   INTEGRATED NAVIGATION AND GUIDANCE SYSTEMS

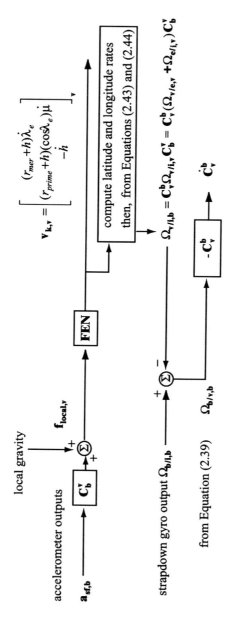

Fig. 10.4 Strapdown DCM mechanization.

NAVIGATION COMPUTER SIMULATIONS

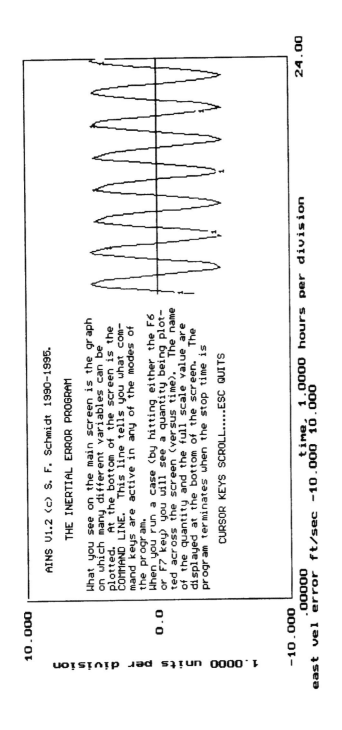

Fig. 10.5 AINS program help (F1 key).

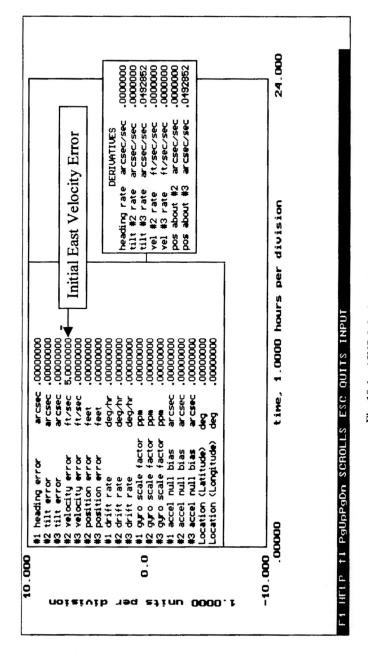

Fig. 10.6 AINS default settings (F4 key).

NAVIGATION COMPUTER SIMULATIONS

Table 10.1 East velocity error effects at different positions on the earth
(Initial east velocity error = 5 ft/s)

Type of error propagation	Latitude and longitude at Palmdale	Longitude of Palmdale with latitude = 0	Latitude of Palmdale with longitude = 0
East velocity, fps			
North velocity, fps		Maximum error/time from start	
East position, ft			
North position, ft	4000 ft/+10 h		
Heading error, arcs			

10.2.2.1 Effect of actual position on error propagation. Begin your investigation by determining the effect of the initial default East velocity error of 5 ft/s on resulting position, velocity, and heading error propagation if the INS is stationary at Palmdale Air Force Plant 42, California, with latitude 34°37.76' North and longitude 118°05.07' West. Fill in the Table 10.1 with your results.

Remember that the Greenwich Meridian choice for zero longitude is totally arbitrary and cannot be deduced from any observation in nature. Latitude, on the contrary, is easily observed in nature by the North Star angular elevation from the horizon. To obtain the data for the table, use the F5 key to enter the latitude and/or longitude of Palmdale at the bottom of the list. Check your plot for North position error against the North position error plot in Fig. 10.7 to see if you are on the right track. Note that initial East velocity error excites North errors even in a stationary INS. The two fundamental error modes that are observable in the plot are the Schuler period of approximately 84 min and the Earth's rotation every 24 h.

It is important to note that two actions are required to change time scales. The first action is the F4 key that sets the abscissa maximum plot time and time increment for screen viewing. The second is the F5 key and PageDown key that set the stop time. Note how the North position error plot is cut short at 32 h as shown in Fig. 10.7. This is because the F4 key set the abscissa at 48 h, but the F5 key set a plot compute stop time at only 32 h. Note also from Fig. 10.7 how the F2 key and up/down arrows selected the North position error variable. Following this, a CR opened the scaling window set at 5000 ft, and finally the A (change attributes) was selected from the bottom menu. This allowed the type of line thickness and color to be selected. Finally, the Esc key selected the interactive program from which the F6 or F7 key may be used to generate the error plot.

10.2.2.2 Effect of gyro drift rate on performance. Estimate the nautical mile per hour performance for an INS stationary at Palmdale, California, for a large number of 8-h test runs using East and North position error plots with the three INS gyros drifting at 0.1 deg/h (one-sigma rms). First hit the F3 key to set the defaults, then use the F5 key to enter the Palmdale latitude and longitude. Then set the default East (#2) velocity error to zero and set 0.1 drift rate rms for each of the three gyros. Use the F2 key to select North position error and scale to plus and minus 200,000 ft for 8-h runs. Use the F7 key to do a Monte Carlo simulation. Check your results with Fig. 10.8 shown here for North position error. Does the

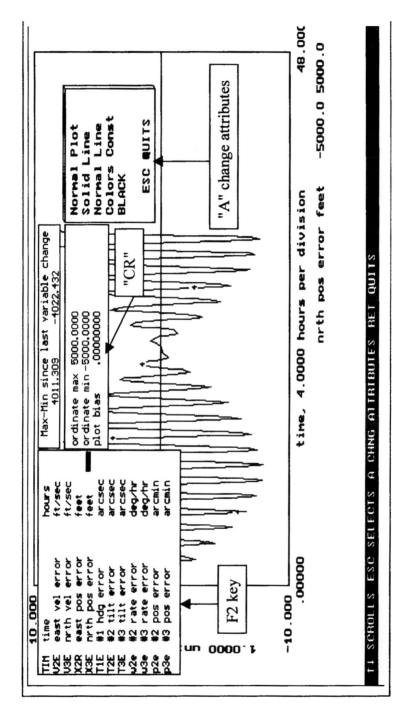

Fig. 10.7 Latitude effect on position error.

NAVIGATION COMPUTER SIMULATIONS

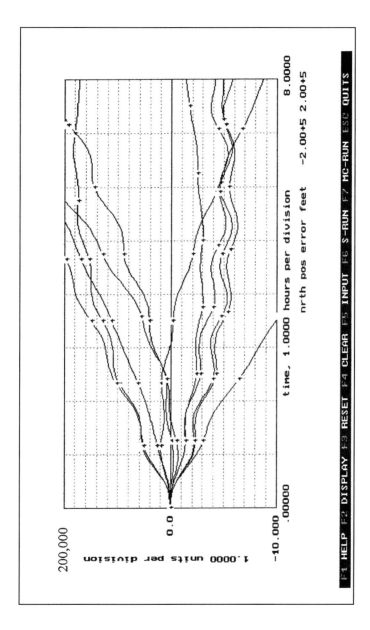

Fig. 10.8   Error due to gyro drift rate.

192   INTEGRATED NAVIGATION AND GUIDANCE SYSTEMS

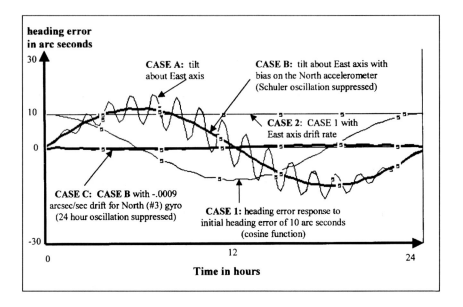

Fig. 10.9 Heading error with different input excitations.

Schuler feedback tend to bound these errors? Repeat selecting the East position error for the plots. What is the estimated nautical mile per hour performance based on these drift errors?

10.2.2.3 Effect of leveling error and gyro drift on heading error.

Plot the heading error performance for an initial 10 arc s heading error. Initialize at Palmdale, California, using F3 to set the defaults and F5 to set East velocity error to zero, then set the heading error to 10 arc s. Use the F2 key to plot the heading error with a scale of plus and minus 30 arc s. Run the simulation by hitting F6 for a deterministic error input and check that a heading error plot results that is a cosine function (CASE 1 in Fig. 10.9) with a maximum value of 10 arc s. Hit the F5 key to obtain the input windows and note that initial heading error causes a tilt rate of 0.0006045 arc s/s about the East (#2) axis. The principle that heading error causes a rotation rate about the East axis is used in gyrocompassing to find true North during the alignment phase.

Now create a drift rate about the East (#2) axis of 0.0006045 arc s/s that should cancel the tilt rate caused by the initial heading error. The plot of heading error should now be a constant 10 arc s of error as shown by CASE 2 in Fig. 10.9. If the plot is a larger cosine function, then check the sign of the drift rate you entered.

Reinitialize the platform using F3 to set the defaults and F5 to set East velocity error to zero, and then use the F5 key to set the leveling (tilt) error about the East (#2) axis to 10 arc s. Note from the input windows that this will cause a heading rate error (-0.0006046 arc s/s) and a North (#3) velocity error rate (0.0004078 arc s/s). Plot the heading error (Case A in Fig. 10.9), then add a North (#3) accelerometer bias of 10 arc s and compare the plots (Case B in Fig. 10.9). Note that the Schuler cycle is suppressed for this combination of inputs. Finally, add a gyro drift about the vertical (#1) of $-.0009$ arc s/s and replot (Case C in Fig. 10.9). The 24-h oscillation now is suppressed as well as the Schuler cycle.

Of course in all of the cases here the INS is not actually flying but is navigating at a stationary position at Palmdale, California, under the influence of initial error sources. In actual flight many or all of the error sources would be excited simultaneously every time the aircraft maneuvered or pulled gs. This is why during flight tests for inertial navigation systems *it is very important to have long flights over specified ground tracks designed to excite and isolate individual error sources* that are driving the INS. Recall from Chapter 9 that navigation system test engineers often plan flight tests that overfly the same path in opposite directions. They do this to keep the INS from suppressing errors along one direction that may occur from a combination of error sources. Time for the legs are scheduled in multiples of a quarter-Schuler-period (approximately 21 min). At least eight test flights are recommended to exercise the INS along all cardinal headings with some level of statistical confidence.

There are other reasons for flying the same ground track during test that will be covered in the KF exercises that follow. It is left to the reader's discretion and initiative to investigate the effects of other error inputs (via the F5 key) into the strapdown INS simulation. Many useful insights are provided in Chapter VII of the AINS book authored by Schmidt that accompanies the interactive exercises.

10.3 AINSBOOK Kalman Filter Simulation

Engage the graphics mode of the KF error propagation simulation from anywhere in the Table of Contents by hitting the F2 key or from the Table of Contents highlight Chapter 6.3 and hit CR. From this point in the manuscript, continue hitting the PageDown key until the interactive KF simulation appears in the graphics mode. You are now ready to complete the laboratory exercises on the AINSBOOK Kalman Filter Simulation. When in the simulation, the screen should display a large, elliptical shape representing the two-dimensional vehicle maneuvering area and two tracking stations at the top-left and bottom-left corners of the maneuvering area.

10.3.1 Two-Dimensional Navigation Problem

The details of the KF implementation and the statistics describing the driving errors for the following exercises are described in Chapter VI of Schmidt's text and are not required to be understood before proceeding with these laboratory exercises. However, a review of Kalman filtering as presented in Chapter 5 in this text may help to understand more fully the concepts presented and emphasized here.

The navigation scenario, described in detail using the F1 help key, is very similar to a vehicle in the horizontal plane using two GPS satellites (SVs) for navigation. A down-looking view of the SVs is shown in Fig. 10.10 with the SVs identified by small circles labeled "1" and "2". The vehicle navigates in two dimensions in the plane of the page. A 16-state EKF is implemented to estimate the vehicle position and velocity based on range and range-rate measurements from the two reference stations (see Chapter VI of Schmidt's AINSBOOK for details). The EKF estimates the vehicle position and velocity from these measurements and

194   INTEGRATED NAVIGATION AND GUIDANCE SYSTEMS

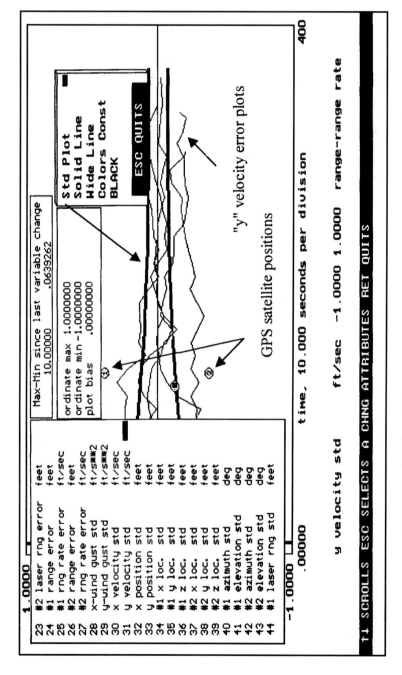

Fig. 10.10 Vehicle tracking on the horizontal plane using two GPS satellites.

from internal models of its motion and error sources, including wind gusts acting on the vehicle. The vehicle is slow moving, starting with a ground speed of 10 fps. Every time the F7 key is pressed, a sample vehicle trajectory is created starting from a random initial velocity direction and stochastic error driving sources.

In Fig. 10.10 the y velocity error has been selected for the plot with default scale limits of plus and minus 1 fps. The error trajectories are created in a Monte Carlo simulation each time the F7 key is pressed. (The F6 key represents a different tracking case study that would not be representative of GPS tracking.) The vehicle is constrained to stay in an elliptical area (not shown in the figure), and the simulation stops when the vehicle hits the edge of the area. Both the actual and estimated positions are shown during the simulation with a view of the horizontal plane from above. A small dot represents the estimated vehicle position, and a small circle represents the actual position. If the circle encloses the dot, then the estimated and actual positions are effectively the same.

In addition to the two-dimensional motion of the dots, error plot realizations of variables selected with the F2 key (as in the INS simulation) are generated vs time each time the F7 key is hit and superimposed on the plot area. This problem is stochastic in nature and is implemented by error sources described by one sigma standard deviations called "std" in the program. The default error sources can be seen using the F5 key and the PageDown command.

The estimated standard deviation of the selected variable (F2 key) from Eq. (4.37) the propagated covariance matrix P (see Sec. 5.2.1.2) also may be plotted vs time. Figure 10.10 shows the window setup needed to plot the standard deviation of y velocity error using the F2 key and subsequent PageDown to variable #31, y velocity std. If the Kalman filter is performing properly, at any given time the standard deviation plot will enclose approximately 67% of the error plot value realizations (assuming that the same scale is used).

The y-velocity error standard deviation is the square root of one of the diagonal elements of P, and the interactive program allows us to plot this element directly from the EKF as shown in Fig. 10.11. Note that successive F7 key hits will generate similar but different y-position standard deviations because of the nonlinear nature of the EKF. This makes the statistics dependent on the trajectory that is followed. Thus it is seen that the covariance elements of P change with time and, for the EKF, with trajectory.

Based on the preceding review of the EKF properties, it is apparent that a similar trajectory should be used when comparing the performance of competing systems. Performance is also dependent on the accuracy of the initial conditions and statistics. If the reference stations represent GPS SVs, these conditions may be impractical to impose for actual testing. With this background, and review of the EKF from Chapter 5, the lab exercises may now be addressed.

10.3.1.1 Divergence of the Kalman filter. The first exercise will demonstrate what happens when the EKF puts too much weight on its own propagated statistics and eventually rejects all measurements, a phenomenon called Kalman filter incest in Sec. 5.3.4 of the navigation text. It is a problem that has challenged researchers since the 1960s and that is demonstrated here by intentionally degrading the accuracy of the error model in the filter. The measurement errors will thus have large covariances relative to the estimated state error covariances.

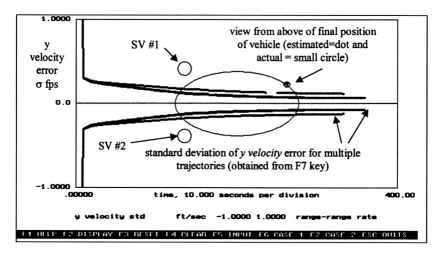

Fig. 10.11 Plotting standard deviation for multiple trajectories.

Use the F5 key to set input locations [1] and [2] (x- and y-wind gust error std) to zero, which eliminates the modeling of wind gust on the vehicle, then PageDown and set [25] and [26] (true values of wind gust std) to 0.02, which increases the true wind gust by an order of magnitude from the default value. Hit the F7 key to generate multiple Monte Carlo test simulation runs. It can be seen in Fig. 10.12 that some of the traces in the x-position error plots diverge. For those divergent traces the estimated position no longer tracks the actual position.

With this setup the KF gains will eventually go to zero [$K_k = P_k(+)H_k^T R_k^{-1}$ and $P_k^{-1}(+) = P_k^{-1}(-) + H_k^T R_k^{-1} H_k$]. Note that the x-position error standard deviation, shown in Fig. 10.12, represents how the EKF thinks it is doing. The underlying cause of this problem is poor modeling of the errors in the KF.

Try to fix this problem by using better modeling and by being more conservative in the wind gust model. Do this by using the F5 key to return to [1] and [2] (x- and y-wind gust error std) and reset them from zero to 0.04 (which is twice the simulation true value that has been set at 0.02). Run at least 10 Monte Carlo runs using the F7 key and plot the x position error for each run (see Fig. 10.13). Use the F2 key to select x-position standard deviation (thick line) as an over-plot and comment on the resulting data plot. Can you deduce from your investigation that conservative modeling (i.e., setting gust rms to 0.04 when true value is 0.02) tends to prevent the divergent behavior by keeping the filter covariance matrix large relative to the measurements? Remember that in real life the true value of the rms wind gust will not be known. Is your one-sigma over-plot reasonable? Does your conclusion hold for y-position error values as well as for the x position errors?

Repeat the preceding analysis for y-velocity error plots and their one-sigma covariance boundary. The state estimate for the propagated covariance error should

NAVIGATION COMPUTER SIMULATIONS 197

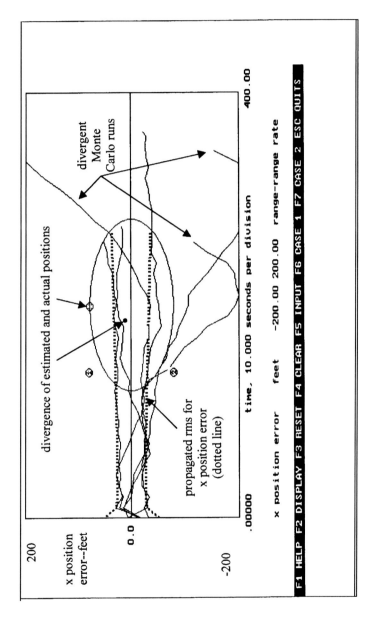

Fig. 10.12   EKF divergence case study.

198   INTEGRATED NAVIGATION AND GUIDANCE SYSTEMS

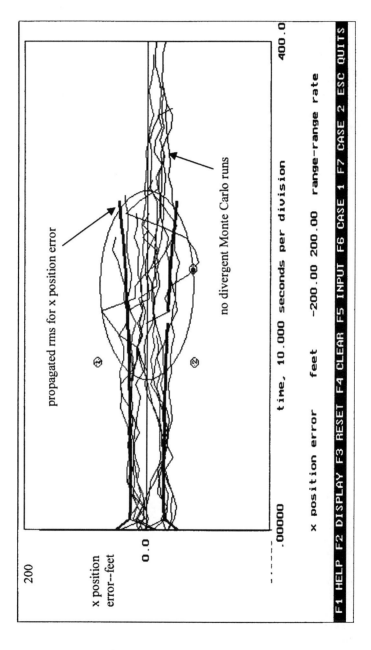

Fig. 10.13 Case study with conservative gust estimates.

now be large enough to contain nearly all of the realizations for all of the time. (Recall that if correct the rms plot would contain approximately 68% of the realizations at any instant of time.) This over-conservative state covariance thus is now large relative to the measurement covariance and allows the measurements to influence the Kalman gains, stopping the divergence.

Remember that should this type of KF divergence be experienced in flight, there will be no warnings or alarm (this is a lack of system integrity). The KF thinks that it is doing great! The propagated error covariance is going to zero in all estimated variables, just as theory guarantees. Your indication of trouble in the cockpit comes when the filter rejects all measurements (even the pilot's updates). If you force it to "swallow" an update, more disconcerting things may occur, one of which is described in the next section.

10.3.1.2 Ramp/reset discontinuity of the Kalman filter.
If the vehicle control system is using the estimated states from a KF, then a discontinuity in the estimate may cause a jump in the control system. Suppose in our example here that the vehicle accepted SV measurement updates every 50 s and used its own internal dynamics model in the interval. This can be set up by first resetting to the default condition using the F3 key, then using the F5 key to set [59] (time update period) to 1 s and [60] (updates per measurement) to 50. Use the F2 key to select [8] (x velocity error) with ordinate max + 1 and plot the x velocity error by running Monte Carlo simulations with the F7 key. Make the x-velocity standard deviation plot using the F2 key to select [30] (x-velocity error std) and overplot as shown in Fig. 10.14 where the discontinuities at every 50-s measurement are clearly seen.

After you are familiar with this scenario, investigate the effect of putting too much weight on the measurements *even when they are valid*. This can be simulated by using the F5 key to set very large values for wind gust uncertainty {set [1] and [2] x- and y-wind gust error std to 0.1}. The model will now not trust itself (the opposite of Kalman filter incest) and at each measurement will yank the estimates back in a sawtooth fashion. Illustrate this by repeating the plots in Fig. 10.14. Change the scale as required but keep it consistent. Figure 10.15 shows this case for y velocity error.

Discuss how this behavior of the filter would manifest itself in the cockpit. What if the error signal drives a component of the flight control system? To say the least, the jolt at ramp-reset time would be disconcerting to crew and passengers! Gradually implementing the correction for control purposes will introduce undesired lag (effective time delay) into the system. Schmidt suggests that for control purposes the negative of the correction be applied at the instant of ramp-reset and that the negative correction be filtered out gradually. In that way the undesired lag is not directly in the control path. Can you think of other implementation schemes?

The navigation algorithm in the preceding scenario is forced to swallow a large update that is inconsistent with its internal model of what is going on. Remember, the underlying problem here is poor performance coupled with a large, forced measurement update. The filter in this case at least knows that it is doing poorly by the increasing size of the propagated covariance matrix. A Monte Carlo analysis would show that the one sigma rms envelope would contain nearly all of the sample realizations.

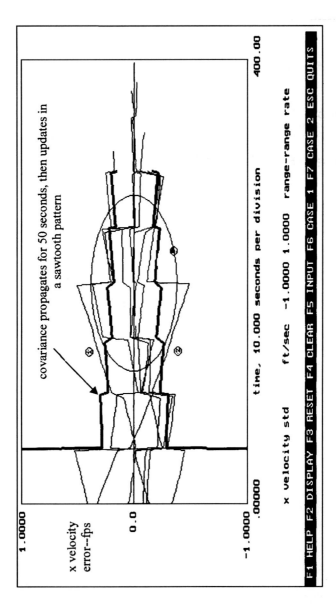

Fig. 10.14　EKF sawtooth error update pattern.

NAVIGATION COMPUTER SIMULATIONS 201

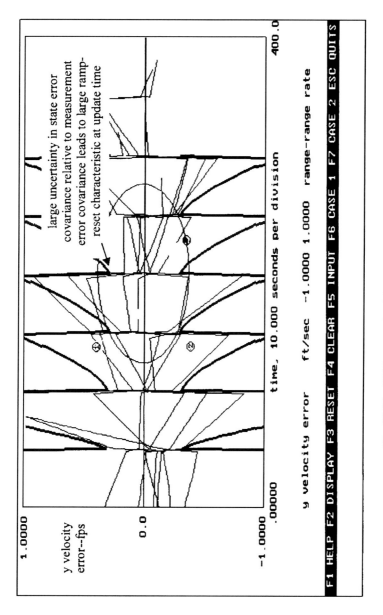

Fig. 10.15   EKF sawtooth with large state uncertainty.

10.3.1.3 Measurement rejection problem.

The preceding exercise showed the difficulty of inserting a needed measurement update after the KF had developed large errors in its internal model. Suppose, however, that the filter was performing well but a large, incorrect measurement occurred to be processed? *If these errors are not spotted and rejected, the navigation errors will grow without bounds.* This is a significant challenge in the development of integrated navigation systems, and it is especially important for updates coming from systems like TACAN where errors in multiples of 40° can occur without warning.

A three-sigma test for measurement rejection was discussed in Sec. 5.3.8 in the navigation text. When the ratio of the measurement error (actual minus predicted measurement error) to the predicted (calculated) measurement error standard deviation is over three, the measurement is rejected. To demonstrate the effect of bad measurements without any rejection, use the F3 key to set defaults, then use the F5 key to set [67] = 10 (this will make 20% of the measurements bad with an error 10 times normal). The effects of this selection on the y-velocity error estimates from the filter are shown in Fig. 10.16. Note that the propagated covariance keeps shrinking and provides little warning of the diverging filter estimates.

Investigate the effect of the three-sigma algorithm on system performance. This can be done by using the F5 key to set [68] = 3 (measurement rejection on with ratio of 3 selected). The plot of Fig. 10.17 shows results with the algorithm "on". You may also investigate the effect of different ratios (set [68] = ratio value) used in the rejection algorithm and the effect of wild outliers (set [67] = maximum value of wild outliers from normal).

Finally, discuss letting the pilot or navigator in the cockpit have an override button so that, if the measurement is known to be good, the filter can be forced to accept it. Contrast this with the capability of selecting the rejection ratio for the Kalman filter. The issues of how much control and the type of control that the flight crew should exercise over an integrated navigation and guidance system are important topics that affect safety of flight as well as situational awareness.

10.3.1.4 Effect of reference station geometry on solution.

If we consider the reference stations as GPS pseudolites located on the ground providing range and range rate information, then the dominant effects of geometry can be investigated by moving the reference station location(s). This is easily done in the simulation using the F5 key to select the #1 reference station position relative to center of the maneuvering area {[49], [50], [51]} and then repeating the process for the #2 reference station position {[52], [53], [54]}. The settings in [49] through [54] thus determine the pseudolite position relative to the user and the user's estimated position (remember that the red dot depicts the estimated position in the simulation).

For example, we know intuitively that if the stations are in a line with the user that geometry will cause a problem. To see this effect in the simulation, first set the defaults with the F3 key, then use the F5 key to set [49] and [52] to zero, which effectively puts the pseudolites directly above and directly below the vehicle. Use the F2 key to select either the x or y position errors and run the Monte Carlo simulation several times using the F7 key. From Fig. 10.18 it can be seen that the EKF is divergent in many cases. Investigate the y-position error plots to verify that the EKF estimates the y coordinate properly. Note that the geometry in this case causes a divergence only in x-position error estimation.

NAVIGATION COMPUTER SIMULATIONS 203

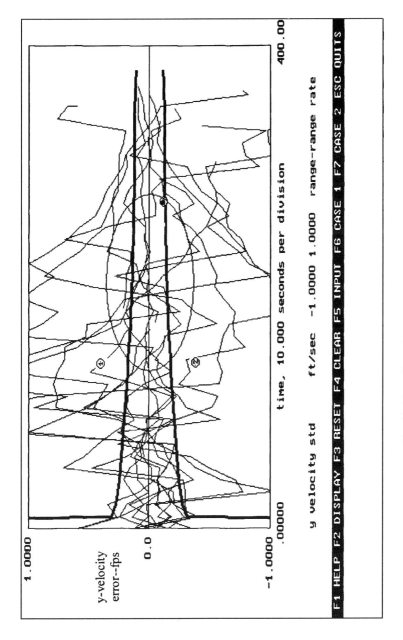

Fig. 10.16 Effect of processing outlier measurements.

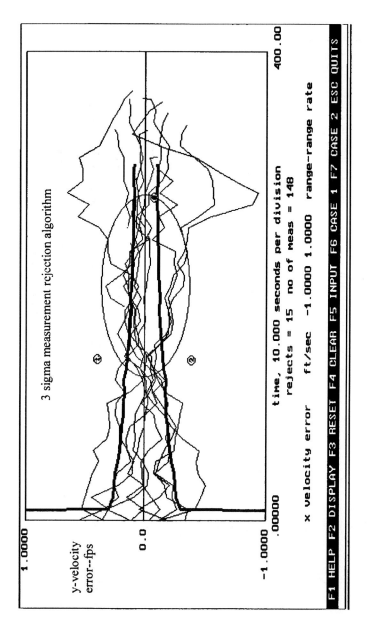

Fig. 10.17 Performance with outlier rejection algorithm.

NAVIGATION COMPUTER SIMULATIONS 205

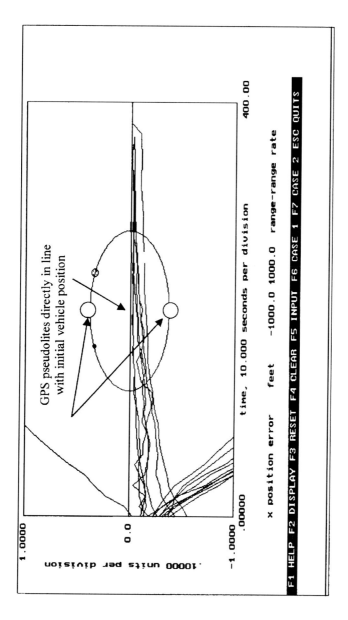

Fig. 10.18 Geometry effects on EKF stability.

Try placing the pseudolites on a horizontal line and see if that causes a divergence in the y position error. The important thing is to recognize the dominant effects of reference station geometry relative to the user. No matter how well-modeled or implemented, even the best EKF cannot compensate for the potential loss of mathematical observability caused by poor geometry. Recall from Chapter 6 a metric called GDOP was defined that measured this loss of precision resulting from poor satellite geometry with respect to the user.

10.3.2 Summary

This completes the laboratory exercises provided with the navigation text, but a wealth of material remains in AINSBOOK that has not been covered here. The important points that these exercises intend to convey are the following:

1) INS systems are very complex and require a computer to simulate their performance. A flying INS simulation would require a trajectory generator whose output would both generate the true trajectory for reference and provide the specific force and rate inputs to the INS simulation. The trajectory must be long enough to identify the Schuler period in the error plots and be repeatable so that error sources can be isolated and identified from test.

2) EKF algorithms have similar performance only when following similar trajectories. Therefore, do not underestimate the effects of poor geometry on the EKF solution. Remember that measurement acceptance/rejection criteria and the sawtooth characteristics of the measurement update process require careful attention to be implemented properly.

The EKF estimates the covariance matrix elements, but this estimate may either diverge without limit or collapse to zero given the proper environment. Recall from the navigation text that if Kalman filters are used to drive Kalman filters (loosely coupled configuration), an unstable covariance solution may bring the entire navigation solution to a halt. Finally, remember that measurements should be independent of each other—it is important to give the EKF a sanity check every once in a while.

10.4 Closure

"Since all our measurements and observations are nothing more than approximations to the truth, the same must be true of all calculations resting upon them, and the highest aim of all computations made concerning concrete phenomena must be to approximate, as nearly as practicable, to the truth."

C. F. Gauss (1809)

The scientific method is largely a give and take between unbiased observations, analyses, and hypothesis testing. Ironically, simulation is often a "reality check" for the empiricist who has devoted his life to making unbiased observations. Galileo, after inventing the telescope in 1610, made spectacular errors as well as spectacular observations. He insisted, for example, that Saturn was actually three spheres because he had observed it! Today we know that what was observed as Saturn's spheres are really its rings. In 1650 a Jesuit astronomer, Giovanni Riccioli, observed that Mizor, the second star in the Big Dipper's handle, was a "double star" system. Today we know that Mizor and its sister Mizor B are double double star systems. Isaac Newton, despite his genius in describing motion, never understood the complicated dynamics of the Earth's moon. Today, mathematical modeling and simulations augment our observations and hypothesis testing and not only help to protect us from spectacular errors, but often provide uncommon insights.

Stanley F. Schmidt, the author of the interactive computer programs in this chapter, is currently a consultant from Palo Alto, California, with a strong educational mission. He is a firm believer in the use of systems simulation to provide educational insight as well as engineering analysis and has written many simulation programs of an educational nature. Two of these have been presented here. His programs excite the interest and technical curiosity of the user, and they represent one of the best educational applications of computers in promoting technical literacy and knowledge in an increasingly complex world.

A
Abbreviations and Acronyms

ADI	attitude director indicator
AFCS	automatic flight control system
AGTFT	antijam global positioning system technology flight test
AHRS	attitude-heading reference system
AIM	Aeronautical Information Manual
AINS	aided inertial navigation system interactive computer exercises
AINSBOOK	aided inertial navigation system computer text by Stanley Schmidt
AR	ambiguity resolution techniques for solving the $N\lambda$ problem
ARSR	air route surveillance radar
ASR	aircraft surveillance radar
Baro	barometric, as in baro-altimeter
BIT/SIT	built-intest/system integrated test
C/A code	coarse acquisition code for standard positioning service that identifies space vehicle and transmits message
CADC	central air data computer
CADGPS	carrier-aided differential global positioning system (same as CDGPS and RTK)
CDGPS	carrier-phase differential global positioning system (same as CDGPS and RTK)
CDU	control display unit
CEP	circular error probable (radius of dispersion enclosing 50% samples)
CFIT	controlled flight into terrain
CIGTF	Central Inertial Guidance Test Facility
CIRIS	competely integrated reference instrumentation system
CR	carriage return key
CTF	Combined Test Force (military)
DCM	direction cosine matrix
DD	double-differencing processing technique
DGPS	differential global positioning system
DH	decision height (feet)
DMA	Defense Mapping Agency
DME	distance measuring equipment
DNSS	Defense Navigation Satellite System
DOP	dilution of precision
DR	dead reckoning
drms	twice the rms radial distance error (95–98% of statistical population)
EADI	electronic attitude director indicator
ECEF	Earth-centered, Earth-fixed

ECM	electronic countermeasures
EFIS	electronic flight instrumentation system
EGI	embedded global positioning system/inertial navigation system
EHSI	electronic HSI
EKF	extended Kalman filter (nonlinear systems)
ESC	escape key
ESG	electrically suspended gyro
F1	keyboard help key in AINSBOOK program
F2	keyboard key to display plot variables in AINSBOOK program
F3	keyboard key to reset default values in AINSBOOK program
F4	keyboard key to select plot parameters in AINSBOOK program
F5	keyboard key to set parameter values in AINSBOOK program
F7	keyboard key for a Monte Carlo simulation run in AINSBOOK program
FAA	Federal Aviation Administration
FEN	fundamental equation of navigation
FLIR	forward-looking infrared
FOG	fiber optic gyro
FRP	Federal Radionavigation Plan
FTE	flight technical error
GDOP	geometric dilution of precision (trace of residual covariance matrix)
GLONASS	Russian Navigation Satellite System
GMT	Greenwich Mean Time (same as Universal Coordinated Time)
GNSS	Global Navigation Satellite System (global positioning system plus GLONASS)
GPS	NAVSTAR global positioning system
GPS IIIA	Military Aircraft Five-Channel GPS Receiver
GPSI	global positioning system/inertial navigation system integration
GPWS	ground proximity warning system
GSS	gyroscope stabilized system
HDIS	high dynamics instrumentation set
HDOP	horizontal dilution of precision (diagonal element of residual covariance)
HIG	hermetic integrating gyro
HOT	higher order terms (as in a mathematical expression)
HUD	head-up display
IB	integrity beacon (low-power ground-based pseudolite)
IBLS	integrity beacon landing system
ICAO	International Civil Aviation Organization
IFR	instrument flight rules
ILS	instrument landing system
IMU	inertial measurement unit
INS	inertial navigation system
IRIG	InterRange Instrumentation Group Time
ISA	integrated accelerometers
JDAM	joint direct attack munition
JPO	Joint Program Office (NAVSTAR global positioning system is a JPO)

APPENDIX A: ACRONYMS 211

KF	Kalman filter (linear systems)
LADGPS	local area differential global positioning system
LAT	latitude
LLINS	local-level inertial navigation system
LNES	linearized navigation equation set for global positioning system
LON	longitude
LORAN	long-range navigation
LOS	line of sight
MAGR	miniature airborne global positioning system receiver
MEM	micromechanical gyro
MFD	multifunction display
MPP	most probable position
NAS	National Airspace System
NAVAID	air navigation radio aid
NAVSTAR	global positioning system
NDB or ADF	nondirectional radio beacon (audio direction finder)
NDI	nondevelopmental item
NED	North-East-Down
NES	navigation equation set for global positioning system
NSTB	national satellite test bed
OCR	operational control segment
OFP	operational flight program
OMEGA	low frequency radio-navigation aid
P code	Precision code (37 weeks long)
PAR	precision approach radar
PD	positive definite matrix property (covariance P must be PD)
pdf	probability density function
PDOP	position dilution of precision (root-sum-square of two covariance diagonal elements)
PLGR	precision light weight global positioning system receiver
PPS	precise positioning service (selective availability off) for global positioning system
PR	pseudorange
PRN	pseudorandom noise (unique pattern for each space vehicle)
PVAT	position, velocity, attitude, time (navigation system outputs)
RADAR	radio detection and ranging
RAIM	receiver (global positioning system) autonomous integrity management
RLG	ring laser gyro
rms	root-mean-square
RNAV	area navigation
RNP	required navigation performance
RSS	root-sum-square
RVR	runway visual range
SA	selective availability (if "ON" degrades the global positioning system signal for standard positioning service)
SA/A	situation awareness/assessment
SD	standard deviation, $\sigma = $ rms
SDS	strapdown inertial navigation system

SEP	spherical error probable
SM	statute mile
SOLGR	special operations lightweight global positioning system receiver
SPO	Systems Program Office
SPS	standard positioning service (selective availability on) for global positioning system
SSINS	space-stabilized inertial navigation system
STD INS	standard inertial navigation system
SV	space vehicle (global positioning system satellite)
TACAN	tactical air navigation
TAI	International Atomic Time
TCH	threshold crossing height (feet)
TDOP	time dilution of precision (diagonal element of residual covariance)
TGAT	tactical global positioning system antijam technology
THAAG	tactical high antijam global positioning system guidance program
TNP	total navigation payload (transmitted by space vehicle)
TOA	time of arrival
TSPI	time-space-position data from a test range
UERE	user-equivalent range error
UT1	Universal Time 1 (time scale based on earth's rotation rate)
UTC	Universal Coordinated Time
UTC(USNO)	Naval Observatory Time
VDOP	vertical dilution of precision (diagonal element of residual covariance)
VFR	visual flight rules
VOR	VHF omnidirectional range
VORTAC	VHF omnidirectional range combined with tactical air navigation at same location
W	watts
WAAS	wide area augmentation system (Federal Aviation Administration Program for differential global positioning system nationwide)
WADGPS	wide area differential global positioning system (basis for wide area augmentation system)
WSMR	White Sands Missile Range, New Mexico
Y code	global positioning system encoding scheme used by the military
Zulu	Greenwich Mean Time (same as Universal Coordinated Time)

B
Discussion Questions

1. No matter how complex integrated navigation systems become, to provide navigational situational awareness in the cockpit their outputs should include
   a) tracking information including angle of attack and sideslip during take-off and climb, position updating during cruise and descent, and course deviation from either an approved ILS or PAR during approach to landing.
   b) radionavigation position fixes and true airspeed in knots with Doppler velocity updates.
   c) position, velocity, and attitude relative to the local Earth's surface and time relative to some common standard (PVAT).
   d) All of the above.
   e) None of the above.

2. What is the primary reason that the vertical channel of an INS requires aiding in-flight and how is this aiding done?
   a) The vertical INS channel requires aiding because of the lag in the vertical velocity. The channel is always aided with GPS altitude.
   b) The differential equation governing the vertical channel is unstable; the channel is normally stabilized with the barometric altimeter.
   c) The vertical channel has a double integration of gravity in the vertical channel that requires gravity to be known exactly, which never occurs. Any long-term stable altitude reference may be used to aid this channel.
   d) Both b) and c) are substantially correct.
   e) None of the answers above are correct; the INS vertical channel does not require aiding.

3. How is it possible to speed up the gyrocompassing alignment of an INS?
   a) It is not possible to speed up this alignment because it totally depends on the rotation rate of the Earth.
   b) If a larger and faster computer is used, the gyrocompassing alignment time can be made to occur even faster than the leveling time for the INS platform.
   c) It is only possible to speed up gyrocompassing alignment time by directly sighting on the North Star during alignment.
   d) If an excellent heading is given to the INS at the start of alignment, you have done the most important thing to speed up alignment time that normally depends on Earth's rotation rate.
   e) If the heading of the INS can be slaved to the directional gyro heading in-flight, gyrocompassing is not important.

4. Geodetic latitude at a point P is the angle between the equatorial plane and
   a) the local vertical (plumb bob) at point P.

b) the perpendicular to the geoid at point P.
c) the normal to the WGS 84 ellipsoid at point P.
d) a line pointed at the North Star.
e) the plane containing the Greenwich Meridian at point P.

5. Which of the following time standards is continuous, that is, does not have discrete jumps programmed into it from time to time:
   a) GPS time.
   b) GMT time.
   c) UTC (USNO) time.
   d) Zulu time.
   e) Both b) and d) are correct.

6. What is gimbal lock and how can it be avoided?
   a) At a pitch angle of plus or minus 90 deg, a platform isolated by three or fewer gimbals from aircraft motion will lock with the mounting case and tumble the gyros. It can be avoided by using four or more gimbals.
   b) At a pitch angle of plus or minus 90 deg, the Euler heading and roll angles are undefined. This may lead to a divide by zero in the computation of heading and roll. Quaternion representation is used to avoid this computation problem in simulations for fighter aircraft.
   c) Both a) and b) are generally true. Gimbal lock may be a physical or a computational problem.
   d) Gimbal lock is caging the attitude gyro after a heavy dose of acrobatic flight. It cannot be avoided unless you fly straight and level.
   e) Gimbal lock is the loss of a tracking loop by the GPS antenna. It can be avoided by having the INS track the GPS satellites in view.

7. A situation occurs in-flight where the integrated INS/GPS system will not automatically accept an update from the pilot that the pilot knows *for certain* is more accurate than the current indicated position.
   a) This is because the Kalman filter has not been tuned recently. Enter the update regardless of the warnings that the systems provide.
   b) This has been designed into the system by engineers who do not trust pilot inputs to the system. Live with the error and do not attempt further updates.
   c) This may be caused by the Kalman filter putting more weight than it should on its own internal model covariance and less weight on external measurements. If you force the filter to accept the full update, it could cause the navigation solution to degrade rapidly. Whatever you decide, have a backup navigation plan!
   d) The Kalman filter has become unstable. Turn all navigation equipment off, including the GPS, and try to realign the INS in-flight.
   e) There is no way that both the INS and the GPS could be wrong at the same time.

8. A Kalman filter weights the accuracy of what two sources of navigation information in determining its own estimate of the navigation solution?
   a) The covariance of the propagated error estimates for position, velocity, attitude, etc., and the covariance of the measurement source(s).

APPENDIX B: DISCUSSION QUESTIONS 215

   b) The accuracy and speed of the computer and the requirements of the real-time cockpit display.
   c) The CEP of the last updated position and the CEP of the next anticipated measurement.
   d) The dead-reckoning solution in the computer and the Doppler radar.
   e) The GPS pseudorange and pseudorange rate and the INS position, velocity, and time.

9. The extended Kalman filter in a tightly coupled configuration of a GPS and INS (GPSI)
   a) provides the most probable position, velocity, attitude, and GPS time.
   b) accepts raw data inputs from both the GPS and the IMU.
   c) Exhibits trajectory dependent but generally stable performance.
   d) Is preferred to the loosely coupled filter-driving-filter configuration that has exhibited stability problems.
   e) All of the above are substantially correct.

10. If the required navigation accuracy for a piloted precision approach using a GPS/INS integrated system is 5 m rms, and you are required to allow 4 m rms for all other error sources, including typical pilot proficiency, what should the rms accuracy of the GPS/INS be?
    a) 3 m rms.
    b) 5 m rms.
    c) 9 m rms.
    d) 16 m rms.
    e) 25 m rms.

11. When using the GPS in-flight, the GDOP tells the pilot that
    a) the GPS may be used without reservation if GDOP is greater than 5.
    b) the number of satellites that the receiver is seeing are equal to the GDOP.
    c) the GPS is suitable for use if GDOP is less than 3.
    d) the precision expected is equal to the GDOP in meters CEP.
    e) the satellite configurations are all broadcasting correct information.

12. GPS was considered a unique use of new technologies because
    a) GPS signals were broadcast by satellites in space, a new idea when it was conceived.
    b) GPS was the first system to provide worldwide navigation coverage regardless of weather.
    c) GPS was the first joint program with the U.S. Navy (TRANSIT), U.S. Army, and the USSR (GLONASS) to be approved by Congress.
    d) GPS used pseudorandom codes for TOA positioning called pseudoranging, spread-spectrum technology for simultaneous, jam-resistant satellite transmissions, depended on ultrareliable timing sources and Earth models.
    e) GPS has no unique technologies—it was simply an evolution of the existing Air Force Program 621, which the Air Force inherited from the Navy.

13. A suitably equipped GPS receiver that can solve the $N\lambda$ ambiguity problem is capable of
    a) legally flying a CAT III approach to automatic touchdown and roll-out provided ATC is notified.

216   INTEGRATED NAVIGATION AND GUIDANCE SYSTEMS

b) position and velocity accuracies better than that of fully capable military P(Y) code systems.
c) accuracies in the centimeter level even in commercial sets not authorized to use the military codes for precise accuracy.
d) Both b) and c) are correct.
e) None of the above are correct.

14. The general idea behind DGPS is to
   a) transmit actual Earth-referenced position and velocity corrections from a reference GPS at a known location to users in the general area.
   b) use a GPS transmitter at a surveyed ground location as a reference for users in the vicinity who can lock-on to its signal much like a transponder.
   c) use a GPS at a surveyed ground station to transmit pseudorange and pseudorange rate corrections for each satellite to users in the general area.
   d) eliminate all atmospheric, multipath, and receiver-based errors in the GPS solution.
   e) use a DD technique to lock onto the L1 carrier signal, providing accuracies at the centimeter level.

15. GPSI, MAGR, and EGI stand for
   a) GPS/INS integration, Miniature airborne GPS receiver, Embedded GPS/INS.
   b) General Purpose Systems Integration, Military Autonomous GPS Receiver, and Elevated GPS integration issues.
   c) GPS/INS integration, Military autonomous GPS Receiver, Embedded gyro insertion.
   d) General purpose systems integration, Military autonomous GPS receiver, and Embedded GPS/INS.
   e) None of the above.

16. GPS and INS are ideal candidates for integration because
   a) GPS outputs position, velocity, and attitude and INS outputs position, velocity, and time.
   b) GPS sets are ridiculously easy to connect to an INS in the loosely coupled configuration with no degradation of system integrity.
   c) GPS is a low output rate (low bandwidth) PVT device with good long-term performance and INS is a high output rate (high bandwidth) PVA with good short-term performance device requiring long-term aiding.
   d) INS is a low output rate (low bandwidth) PVT device with good long-term performance and GPS is a high output rate (high bandwidth) PVA with good short-term performance device requiring long-term aiding.
   e) Both a) and d) are correct.

17. GPS implements systems integrity by:
   a) Solving this problem with the $N\lambda$ solution.
   b) No GPS satellite has ever put out bad data; the integrity problem is non-existent.
   c) Receiver autonomous integrity management (RAIM).
   d) All of the above are correct.
   e) None of the above are correct.

APPENDIX B: DISCUSSION QUESTIONS   217

18. The purpose of every INS is to use sensors, a platform mechanization, and a computer to
   a) produce acceleration and velocity with respect to the Earth's surface, which may be integrated for position, and to provide attitude with respect to the vehicle-carried-vertical frame.
   b) maintain a stable platform for the sensors, especially the gyros, especially if ring laser gyros are employed.
   c) solve the frequently used equation of navigation (FUEN) and display results to the pilot.
   d) generate direct-computer matrices (DCM) so that attitude with respect to the Earth-centered, Earth-fixed reference frame is available for display to the cockpit.
   e) implement a Kalman filter so that measurement updates can be made from the GPS and ADF.

19. It is important to remember the following about Kalman filters while you are flying and navigating an aircraft with an INS:
   a) Their performance is trajectory dependent.
   b) They may diverge (lock-up) by refusing to accept any measurement updates from any source, or they may diverge if forced to accept too big an update, no matter how accurate.
   c) They generate estimates of position and velocity accuracy used in the update process.
   d) When working properly, the error time plots follow a sawtooth shape with discontinuities at the measurement update time.
   e) All of the above.

20. If competing INS systems are being flight tested, it is important that
   a) the flight tests have a repeatable trajectory.
   b) an EKF and GPS be operating on the test vehicle.
   c) only nonmaneuvering profiles be flown.
   d) the flight test legs be multiples of 8.4 h.
   e) None of the above.

C
Web Sites by Chapter

URL Sites Related to Chapter 1

http://www.aero.org/publications/GPSPRIMER/WhatisNav.html
gopher://venus.hyperk.com:2101/
http://www.navcen.uscg.mil/
http://daniel.calpoly.edu/~dbiezad/

URL Sites Related to Chapter 2

http://yi.com/home/BomannsAlfred/gc.htm
http://164.214.2.59/geospatial/products/GandG/geolay/toc.htm
http://www.utexas.edu/depts/grg/gcraft/notes/datum/datum.html
http://www.timing.se/
http://tycho.usno.navy.mil/
http://www-nmd.usgs.gov/
http://acc.nos.noaa.gov/
http://164.214.2.59   {NIMA}
http://164.214.2.59/GandG/pubs.html

URL Sites Related to Chapter 3

http://www.tspi.elan.af.mil/ins.html
http://www.wlmn.eglin.af.mil/public/mnag/lowcost.html
http://www.wlmn.eglin.af.mil/public/mnag/its.html
http://daniel.calpoly.edu/~dfrc/World/

URL Sites Related to Chapter 4

http://home.cern.ch/asdoc/shortwrupsdir/e230/top.html
http://www-math.mit.edu/~gs/books/books.html

URL Sites Related to Chapter 5

http://www.tspi.elan.af.mil/kalman.html
http://www.cs.unc.edu/~welch/kalmanLinks.html
http://www.cs.unc.edu/~welch/kalmanBooks.html
http://www.cs.unc.edu/~welch/kalman/kalman.html

URL Sites Related to Chapter 6

http://www.laafb.af.mil/SMC/CZ/homepage/index.html
http://www.op.dlr.de/~igex98op/monitor/monitor.htm

http://www.ion.org/
http://www.sni.net/oriondc/
http://www.utexas.edu/depts/grg/gcraft/notes/gps/gps.html
http://www.sdd.sri.com/GPS/gps.html
http://www.navtechgps.com/
http://www.sni.net/kawin/pages/sawhatis.htm
http://tycho.usno.navy.mil/gps.html
http://www.trimble.com/gps/fsections/aa_f1.htm
http://www.rssi.ru/SFCSIC/glonass.html
http://satnav.atc.ll.mit.edu/
http://satnav.atc.ll.mit.edu/java/SVOverlay/SVOverlay.shtml
http://www.starlinkdgps.com/start.htm
http://www.aero.org/publications/GPSPRIMER/index.html
http://www.mapmania.com/

URL Sites Related to Chapter 7

http://www.unidata.ucar.edu/SuomiNet/index.html
http://www.csr.utexas.edu/texas_pwv/
http://www.stanford.edu/group/GPS/Projects/WAAS/
http://einstein.stanford.edu/gps/gps_pubs.html

URL Sites Related to Chapter 8

http://www.gpsworld.com/
http://www.tspi.elan.af.mil/gps.html
http://www.cnde.iastate.edu/staff/swormley/gps/satellites.html

URL Sites Related to Chapter 9

http://www.acq.osd.mil/te/pubfac/holloman.html
http://www.wlmn.eglin.af.mil/public/mnag/mmimu.html
http://www.wlmn.eglin.af.mil/public/weapflgt.html
http://www.dfrc.nasa.gov/
http://FlightTest.navair.navy.mil/

URL Sites Related to Chapter 10

http://www.tspi.elan.af.mil/kalman.html
http://www.gpsoftnav.com/
http://www.sni.net/oriondc/

References

"ASCC Air Standard 53/11B," U.S. Air Force, August 15, 1968.

"6521 Range Capabilities Handbook," Air Force Flight Test Center, Edwards Air Force Base, June 15, 1968.

"Air Navigation," U.S. Air Force, AFM 51-40, March 15, 1983.

"GPS NAVSTAR User's Overview," ARINC Research Corporation, Los Angeles, YEE-82-009D, March, 1991.

"Global Positioning System, NAVIGATION Vol. IV," Institute of Navigation, Alexandria, Virginia, 0-936406-03-8, 1993.

"NAVSTAR GPS User Equipment Novella," GPS Joint Program Office, June 30, 1996.

Global Satellite Navigation System User's Guide, Boulder, Colorado, Orion Dynamics and Control, Inc., 1997.

Abusail, P. A. M., Tapley, B. D., and Schutz, B. E., "Autonomous Navigation of Global Positioning System Satellites Using Cross-Link Measurements," *AIAA Journal of Guidance, Control, and Dynamics*, Vol. 21, No. 2, pp. 321–327, March–April, 1998.

Axelrad, P., Comp, C. J., and MacDoran, P. F., "SNR Based Multipath Error Correction for GPS Differential Phase," *IEEE Transactions on Aerospace and Electronics Systems*, Vol. 32, No. 2, pp. 650–660, 1997.

Barrows, A., Enge, P., Parkinson, B., and Powell, D., "Evaluation of a Perspective-View Cockpit Display for General Aviation Using GPS," *Journal of the Institute of Navigation*, Vol. 43, No. 1, pp. 55–70, Spring, 1996.

Bierman, G. J., and Thornton, C. L., "Numerical Comparison of Kalman Filter Algorithms," *Automatica*, Vol. 13, pp. 23–35, 1977.

Biezad, D. J., "Normalized Predictive Deconvolution: Multichannel Time-Series Applications to Human Dynamics," in *Control and Dynamic Systems*, Vol. 31, C. T. Leondes, Ed. San Diego: Academic Press, 1989, pp. 193–257.

Bowditch, N., *American Practical Navigator*, Washington D.C., U.S. Government Printing Office, 1962.

Brainard, J., "An Ever Taller Everest?," *Science News*, Vol. 153, No. 26, p. 410, June 27, 1998.

Britting, K. R., *Inertial Navigation Systems Analysis*, Massachusetts Institute of Technology, Measurement Systems Laboratory, 1971.

Brown, R. G., and Hwang, P. Y. C., *Introduction to Random Signals and Applied Kalman Filtering: With Matlab Exercises and Solutions*, 3rd ed., New York, John Wiley and Sons, Inc., 1996.

Brown, R. L., "Home Off the Range: A Portable Military Ground-Truth System," *GPS World*, Vol. 8, No. 6, pp. 29–38, June, 1997.

Broxmeyer, C., *Inertial Navigation Systems*, McGraw-Hill, 1964.

Carlson, N. A., and Berarducce, M. P., "Federated Kalman Filter Simulation Results," *Journal of the Institute of Navigation*, Vol. 41, No. 3, pp. 297–322, Fall, 1994.

Ceva, J., Bertiger, W., Muellerschoen, R., Yunch, T., and Parkinson, B., "Incorporation of Orbital Dynamics to Improve Wide-Area Differential GPS," *Journal of the Institute of Navigation*, Vol. 44, No. 2, pp. 171–180, Summer, 1997.

Cobb, S., and O'Connor, M., "Innovation: Pseudolites–Enhancing GPS with Ground-Based Transmitters," *GPS World*, Vol. 9, No. 3, pp. 55–60, March, 1998.

Coffee, J., and Maganty, P., "An Integrated DGPS/INS Navigation System for a Ballistic Missile: Design and Flight Test Results," *Journal of the Institute of Navigation*, Vol. 43, No. 3, pp. 273–294, Fall, 1996.

Cohen, C. E., Parkinson, B. W., and McNally, D. B., "Flight Tests of Attitude Determination Using GPS," *Journal of the Institute of Navigation*, Vol. 41, No. 1, pp. 83–98, Spring, 1994.

Cohen, C. E., Pervan, B. S., Lawrence, D. G., Cobb, H. S., Powell, J. D., and Parkinson, B. W., "Real-Time Flight Testing Using Integrity Beacons for GPS Category III Precision Landing," *Journal of the Institute of Navigation*, Vol. 41, No. 2, pp. 145–159, Summer, 1994.

Cohen, C. E., Cobb, H. S., Lawrence, D. G., Pervan, B. S., Powell, J. D., Parkinson, B. W., Aubrey, G. J., Loewe, W., Ormiston, D., McNally, B. D., Kaufmann, D. N., Wullschleger, V., and Swinder, R. J. J., "Autolanding a 737 Using GPS Integrity Beacons," *Journal of the Institute of Navigation*, Vol. 42, No. 3, pp. 467–486, Summer, 1995.

Conway, A., Montgomery, P., Rock, S., Cannon, R., and Parkinson, B., "A New Motion-Based Algorithm for GPS Attitude Integer Resolution," *Journal of the Institute of Navigation*, Vol. 43, No. 2, pp. 179–190, Summer, 1996.

Cook, G., "Glonass Performance, 1995–1997, and GPS-GLONASS Interoperability Issues," *Journal of the Institute of Navigation*, 44, No. 3, pp. 291–300, Fall, 1997.

Cox, D. B., Jr., "Integration of GPS with Inertial Navigation Systems," *Journal of the Institute of Navigation*, Vol. 25, No. 2, pp. 236–245, Spring, 1978.

Crassidis, J. L., and Markley, F. L., "New Algorithm for Attitude Determination Using Global Positioning System Signals," *AIAA Journal of Guidance, Control, and Dynamics*, Vol. 20, No. 5, pp. 891–896, September–October, 1997.

Crum, J., and Smetek, R., "An Overview of GPS Master Control Station Anomaly Detection and Resolution Techniques," *Journal of the Institute of Navigation*, 44, No. 2, pp. 133–152, Summer, 1997.

Dahlen, N., Caylor, T., and Goldner, E., "Tightly Coupled IFOG-Based GPS Guidance Package," *Journal of the Institute of Navigation*, Vol. 43, No. 3, pp. 257–272, Fall, 1996.

Draper, C. S., Wrigley, W., Hovorka, J., *Inertial Guidance*, Pergamon Press, 1960.

Draper, C. S., "On Course to Modern Guidance," *Astronautics and Aeronautics Magazine*, pp. 56–62, February, 1980.

Etkin, B., and Reid, L. D., *Dynamics of Flight: Stability and Control*, New York, Wiley, 1996.

Farrell, J. L., *Integrated Aircraft Navigation*, New York, Academic Press, 1976.

Farrell, J., Djodat, M., Barth, M., and Grewal, M., "Latency Compensation for Differential GPS," *Journal of the Institute of Navigation*, Vol. 44, No. 1, p. 99, Spring, 1997.

Fernandez, M., and Macomber, G. R., *Inertial Guidance Engineering*, Englewood Cliffs, New Jersey, Prentice-Hall, 1962.

Gai, E., "Guidance, Navigation, and Control from Instrumentation to Information Management," *AIAA Journal of Guidance, Control, and Dynamics*, Vol. 19, No. 1, pp. 10–14, January–February, 1996.

Gelb, A., *Applied Optimal Estimation*, Analytic Sciences Corporation, 1974.

Grant, J., "War Games: Troops Train with GPS-Enabled Battlefield Simulation," *GPS World*, Vol. 8, No. 11, pp. 22–31, November, 1997.

Greenspan, R. L., "Inertial Navigation Technology from 1970–1995," *Journal of the Institute of Navigation*, Vol. 42, No. 1, pp. 165–186, 1995.

Gregorius, T., and Blewitt, G., "The Effect of Weather Fronts on GPS Measurements," *GPS World*, Vol. 9, No. 5, pp. 52–60, May, 1998.

Griffiths, B. E., and Geyer, E. M., "Interfacing Kalman filters with the Standard INS," Analytic Sciences Corporation, AFWAL-TR-84-1139, September, 1984.

Haering, E. A., "Booming Blackbird: GPS Goes Supersonic," *GPS World*, Vol. 9, No. 5, pp. 24–32, May, 1998.

Han, S., and Rizos, C., "Comparing GPS Ambiguity Resolution Techniques," *GPS World*, Vol. 8, No. 10, pp. 54–61, October, 1997.

Hansen, A., "Innovation: The NSTB–A Stepping Stone to WAAS," *GPS World*, pp. 73–77, June, 1998.

Huddle, J. R., "Inertial Navigation System Error-Model Considerations in Kalman Filtering Applications," in *Control and Dynamic Systems*, C. T. Leondes, Ed. New York: Academic Press, 1983, pp. 293–339.

Hwang, P. Y. C., "Kinematic GPS for Differential Positioning: Resolving Integer Ambiguities on the Fly," *Journal of the Institute of Navigation*, Vol. 38, No. 1, pp. 1–16, Spring, 1991.

Johnson, A., "GPS Rollover: Compliance Efforts Under Way," *GPS World*, Vol. 8, No. 9, pp. 62–64, September, 1997.

Kailath, T., "An Innovations Approach to Least Squares Estimation," *IEEE Transactions on Automatic Control*, AC-13, No. 6, December, 1968.

Kalman, R. E., "A New Approach to Linear Filtering and Prediction Problems," *ASME Transactions: The Journal of Basic Engineering*, Vol. 82, pp. 35–45, March, 1960.

Kaminski, P. G., Bryson, A. E., and Schmidt, S. F., "Discrete Square Root Filtering: A Survey of Current Techniques," *IEEE Transactions on Automatic Control*, AC-16, pp. 727–733, December, 1972.

Kaplan, E. D., *Understanding GPS Principles and Applications*, Artech House, 1996.

Kayton, M., and Fried, W., *Avionics Navigation Systems*, Wiley, 1969.

Kayton, M., "Navigation: Land, Sea, Air, and Space," *Selected Reprint Series*, L. Shaw, Ed. New York, IEEE Press, 1990.

Kee, C., Parkinson, B. W., and Axelrad, P., "Wide Area Differential GPS," *Journal of the Institute of Navigation*, Vol. 38, No. 2, pp. 123–146, Summer, 1991.

Kelly, R. J., and Davis, J. M., "Required Navigation Performance for Precision Approach and Landing," *Journal of the Institute of Navigation*, Vol. 41, No. 1, pp. 1–30, Spring, 1994.

Kremer, G. T., Kalafus, R. M., Loomis, P. V. W., and Reynolds, J. C., "The Effect of Selective Availability on Differential GPS Corrections," *Journal of the Institute of Navigation*, Vol. 37, No. 1, pp. 40–52, Spring, 1990.

Kwakernaak, H., and Sivan, R., *Linear Optimal Control Systems*, New York, Wiley, 1972.

Langley, R. B., "GLONASS: Review and Update," *GPS World*, Vol. 8, No. 7, pp. 46–51, July, 1997.

Langley, R. B., "The UTM Grid System," *GPS World*, Vol. 9, No. 2, pp. 46–50, February, 1998.

Lapucha, D., Barker, R., and Liu, Z., "High-Rate Precise Real-Time Positioning Using Differential Carrier Phase," *Journal of the Institute of Navigation*, Vol. 43, No. 3, pp. 295–306, Fall, 1996.

REFERENCES

Levy, L. J., "The Kalman Filter: Navigation's Integration Workhorse," *GPS World*, Vol. 8, No. 9, pp. 65–71, September, 1997.

Lin, C. F., *Modern Navigation, Guidance, and Control Processing*, Englewood Cliffs, New Jersey, Prentice-Hall, 1991.

Luenberger, D. G., "Observers for Multivariable Systems," *IEEE Transactions on Automatic Control*, AC-11, No. 2, April, 1966.

Mark, J., Tazartes, D., Fidric, B., and Cordova, A., "A Rate Integrating Fiber Optic Gyro," *Journal of the Institute of Navigation*, Vol. 38, No. 4, pp. 341–354, Winter, 1991.

Masson, A., Burtin, D., and Sebe, M., "Kinematic DGPS and INS Hybridization for Precise Trajectory Determination," *Journal of the Institute of Navigation*, 44, No. 3, pp. 313–322, Fall, 1997.

Maybeck, P. S., *Stochastic Models, Estimation, and Control*, Vol. 1, New York, Academic Press, 1979.

Merhav, S., *Aerospace Sensor Systems and Applications*, New York, Springer-Verlag, 1996.

Miller, A. R., Moskowitz, I. S., and Simmen, J., "Traveling on the Curved Earth," *Journal of the Institute of Navigation*, Vol. 38, No. 1, pp. 71–78, Spring, 1991.

Morf, M., and Kailath, T., "Square-Root Algorithms for Least-Squares Estimation," *IEEE Transactions on Automatic Control*, AC-20, No. 4, pp. 487–497, 1975.

Nikiforov, I. V., "New Optimal Approach to Global Positioning System/Differential Global Positioning System Integrity Monitoring," *AIAA Journal of Guidance, Control, and Dynamics*, Vol. 19, No. 5, pp. 1023–1033, September–October, 1996.

Nordwall, B. D., "GPS Challenges," *Aviation Week and Space Technology*, Vol. 147, No. 22, pp. 58–68, December 1, 1997.

Papoulis, A., *Probability, Random Variables, and Stochastic Processes*, New York, McGraw-Hill, 1984.

Park, C., Kim, I., Jee, G., and Lee, J. G., "Relationships Between Positioning Error Measures in Global Positioning System," *AIAA Journal of Guidance, Control, and Dynamics*, Vol. 20, No. 5, pp. 1045–1047, September–October, 1997.

Parkinson, B. W., and Axelrad, P., "Autonomous GPS Integrity Monitoring Using the Pseudorange Residual," *Journal of the Institute of Navigation*, Vol. 35, No. 2, pp. 255–274, Spring, 1988.

Parkinson, B., Stansell, T., Beard, R., and Gromov, K., "A History of Satellite Navigation," *Journal of the Institute of Navigation*, Vol. 42, No. 1, pp. 109–164, Spring, 1995.

Parkinson, B. W., and Spilker, J. J., "Global Positioning System: Theory and Applications," *AIAA Volume 163 Progress in Astronautics and Aeronautics Volumes 1 and 2*, AIAA, 1996.

Parkinson, B. W., "Origins, Evolution, and Future of Satellite Navigation," *AIAA Journal of Guidance, Control, and Dynamics*, Vol. 20, No. 1, pp. 11–25, January–February, 1997.

Parvin, R. H., *Inertial Navigation*, Princeton, New Jersey, Van Nostrand, 1962.

Pervan, B. S., Cohen, C. E., and Parkinson, B. W., "Integrity Monitoring for Precision Approach Using Kinematic GPS and a Ground-Based Pseudolite," *Journal of the Institute of Navigation*, Vol. 41, No. 2, pp. 159–174, Summer, 1994.

Pervan, B. S., and Parkinson, B. W., "Cycle Ambiguity Estimation for Aircraft Precision Landing Using the Global Positioning System," *Journal of Guidance, Control, and Dynamics*, Vol. 20, No. 4, pp. 681–689, July–August, 1997.

Pinson, J. C., "Inertial Guidance for Cruise Missiles," in *Guidance and Control of Aerospace Vehicles*, C. T. Leondes, Ed. New York: McGraw-Hill, 1963, pp. 113–187.

REFERENCES

Pitman, G. R., *Inertial Guidance*, New York, Wiley, 1962.

Post, E. J., "Sagnac Effect," *Reviews in Modern Physics*, 39, No. 2, pp. 475–493, April, 1967.

Rao, J., "Stellar Body Double," *Natural History*, Vol. 107, No. 4, p. 76, May, 1998.

Renshaw, C., "Moving Clocks, Reference Frames and the Twin Paradox," *IEEE AES Systems*, pp. 27–32, January, 1995.

Rich, B. R., and Janos, L., *Skunk Works*, Little, Brown and Company, 1994.

Robinson, E., *Multichannel Time Series Analysis*, Goose Pond Press, 1983.

Romrell, G., Johnson, G., Brown, R., and Kaufmann, D., "DGPS Category IIB Feasibility Demonstration Landing systems with Flight Test Results," *Journal of the Institute of Navigation*, Vol. 43, No. 2, pp. 131–148, Summer, 1996.

Savage, P., "Strapdown Sensors," AGARD Lecture Series No. 95, 1978.

Savage, P. G., "Strapdown Inertial Naviagtion Integration Algorithm Design Part 2: Velocity and Position Algorithms," *AIAA Journal of Guidance, Control, and Dynamics*, Vol. 21, No. 2, pp. 208–221, March–April, 1998.

Savage, P. G., "Strapdown Inertial Navigation Algorithm Design Part 1: Attitude Algorithms," *AIAA Journal of Guidance, Control, and Dynamics*, Vol. 21, No. 1, pp. 19–28, January–February, 1998.

Schmidt, S. F., "Application of State-Space Methods to Navigation Problems," in *Advances in Control Systems*, Vol. 3. New York: Academic Press, 1966, pp. 293–340.

Schmidt, S. F., "Computational Techniques in Kalman Filtering," London, AGARDograph 139 Theory and Applications of Kalman Filtering, 1970.

Schmidt, S. F., *Aided Inertial Navigation Systems*, Analytical Mechanics Associates, Inc., 1995.

Sciegienny, J., Nurse, R., Wexler, J., and Kampion, P., "Inertial Navigation System Standardized Software Development, Vol. II, INS Survey and Analytical Development," The Charles Stark Draper Laboratory, Inc., Cambridge, Massachusetts, June, 1976.

Siouris, G. M., *Aerospace Avionics Systems: A Modern Synthesis*, Academic Press, 1993.

Siouris, G. M., *Optimal Control and Estimation Theory*, New York, Wiley, 1996.

Skolnik, M. I., "Fifty Years of Radar," *Proceedings of the IEEE*, 73, No. 2, February, 1985.

Sobel, D., *Longitude*, Walker and Company, 1995.

Sofair, I., "Improved Method for Calculating Exact Geodetic Latitude and Altitude," *AIAA Journal of Guidance, Control, and Dynamics*, Vol. 20, No. 4, pp. 824–826, July–August, 1997.

Spence, C. F., and Editor, "Aeronautical Information Manual," McGraw-Hill, 1998.

Spitzer, C., *Digital Avionic Systems*, Englewood Cliffs, New Jersey, Prentice-Hall, 1987.

Strang, G., *Linear Algebra and its Applications*, New York, Academic Press, 1980.

Strang, G., and Borre, K., *Linear Algebra, Geodesy, and GPS*, Wellesley, Massachusetts, Wellesley-Cambridge Press, 1997.

Teague, E. H., How, J. P., and Parkinson, B. W., "Control of Flexible Structures Using GPS: Methods and Experimental Results," *AIAA Journal of Guidance, Control, and Dynamics*, Vol. 21, No. 5, pp. 673–683, September–October, 1998.

van Diggelen, F., "Innovation: GPS Accuracy–Lies, Damn Lies, and Statistics," *GPS World*, Vol. 9, No. 1, pp. 41–45, January, 1998.

Walsh, J., and Wojciech, J. "TCAS in the 1990's," *Journal of the Institute of Navigation*, Vol. 38, No. 4, pp. 383–398, Winter, 1991.

Weill, L. R., "Conquering Multipath: The GPS Accuracy Battle," *GPS World*, Vol. 8, No. 4, pp. 59–64, April, 1997.

Weiss, P., "Ultracold Atoms: New Gravity Yardstick?," *Science News*, Vol. 154, No. 6, p. 87, August 8, 1998.

Widnall, W. S., and Grundy, P. A., "Inertial Navigation System Error Models," Intermetrics, Inc., TR-03-73, May 11, 1973.

Wrigley, W., Hollister, W. M., and Denhard, W. G., *Gyroscopic Theory, Design, and Instrumentation*, MIT Press, 1969.

Index

A

Accelerometers
　errors, 46-48
　pendulous, 45-46
　PIGA, 46
　SFIR, 46
Accuracy, for landing aircraft, 157-158
AINSBOOK
　installation, 181-182
　INS simulation, 185
　Kalman filter simulation, 193
Attitude control, 160
ADI, 10-11
AHRS, 70
Applied specific acceleration, 45
Artificial horizon, See ADI
Attitude, See Euler angles
Angular momentum, 29
Autocorrelation function, See Stochastic process
Average. See expected value
Avionics, 13

B

Baro altitude. See Inertial platform vertical channel
Basis vector, 16-17
BATH alignment, See Inertial platform alignment
Bayes Theorem, 86
Bearing, 4
　beacon (NDB, ADF), 6,8
Boeing, 125, 160
Britting K., 79

C

Central Limit Theorem, 86
Circular error probable. See statistics
Cooper-Harper ratings, 170
Coriolis Theorem, 18-19
Covariance matrix, See Stochastic process
Covariance propagation, 90-91, 101
Craft rate, 3, 34
Cross-correlation function, 89
Cross-variance matrix, 90

D

Datum conversions, 36-37
Dead reckoning, 3-4
Differential GPS, 143-144
　carrier-phase tracking. See Kinematic GPS
　LADGPS, 144, 161
　landing aircraft, 159, 161
　national satellite test bed, 148
　nationwide differential system, 146
　WAAS, 141-147
Direction cosine matrix, 17
Direction cosine matrix, rate of change, 33
Display
　partial flight, 12
　primary flight, 11
Distance radial mean-square, 84
Dithering, 53
DME, 7
Doolittle, Jimmy, 13, 178
Double-differencing technique, 152-155
Draper, Charles, 71, 178-179

E

EADI, 10-11
Earth
    equatorial radius, 21
    flattening, 21
    meridional radius, 23, 55
    polar radius, 21
    prime radius, 23, 55
    WGS84, 21
EFIS, 12
EGNOS, 136-137
EHSI, 11-12
Ellipticity. See earth, flattening
Equinox, 17
Error budget, 74-75, 106
ESG, 50
Euler angles, 25-27
Expected value, 85
Extended Kalman filter. See Kalman filter, EKF

F

FAA, 6
Fiber-optic gyros, 53
Flight management system, 13
Flight technical error, 158
Force
    local, 29-30, 54
    specific, 43, 45
Fundamental equation of navigation, 29

G

Galilean. See inertial frame
Galileo, 206
Gauss, C. F., 206
Gaussian normal distribution, 85-86
GDOP, 93, 112, 122, 134
Geodesy. See Earth
Geoid, 23-25
Geometric mean, 171
Gimbal lock, 27, 58-59
GLONASS, 125, 135-136
g-meter, 44-45
GM/RMS method. See statistics

GNSS, 135, 138
GPS
    accuracy, 112, 146
    aiding INS, 107, 141-143
    availability, 110, 146, 159
    background, 110
    C/A coarse acquisition code, 117, 123
    control segment, 126
    differential. see Differential GPS
    embedded GPS/INS, 143
    errors, 134-135
    hand-off word (HOW), 124
    high dynamic instrumentation set, 162
    instrument approach, 128-129, 139
    integrity, 101, 141, 158
    jammer-to-signal ratio, 128
    jamming susceptibility, 131, 176
    linearization, 116, 119-121
    L5 signal, 125
    navigation message, 125
    navigation payload, 123, 126
    P code: precise positioning service, 112, 123-124
    pseudorange measurements, 35-36, 117-118
    satellite geometry, 122
    satellites Block IIR, IIF, 122-123
    space segment, 122-123
    spread spectrum, 123-124
    standard positioning service, 112
    time bias, 113, 117-118
    transmitted signals L1 and L2, 123-124
    user equipment error, 114, 134
    user segment, 126-128
    Y-encrypted code, 123
Gravity
    anomaly, 20
    plumb-bob, 3
    deflection of vertical, 20, 23, 82
    deviation of normal, 21-23
    local, 20-21, 23
Gravitation, mass, 20, 23-24
Great circle, 4
Ground proximity warning systems, 148
Guidance, 1

Gyro
  apparent precession, 42, 50
  directional, 42-43
  drift, 42, 51, 78, 90, 189, 192-193
  dynamic equation for, 49
  errors, 50-51
  ESG, 50
  Fiber-optic gyro, 53, 177-178
  HIG, 48
  Multi-oscillator RLG, 53
  principles, 40
  ring laser, 51-53
  single DOF, 42, 48-49
  two DOF, 42, 44, 49-50
Gyrocompass 44, 65

H
HDOP, 111, 122
Honeywell, 52
HSI, 11-12
HUD, 12

I
ILS, 7-8
  localizer, 8
  glide-slope, 9
Inertial navigation
  aided INS, 105
  analytic strapdown, 39, 65-67, 183-186
  error equations, 76-82
  flight testing, 167-168, 193
  geometric space-stabilized, 39, 55-56
  local-level, semi-analytic, 56-59
  measurement updates, 140
Inertial platform
  alignment, 63-65
  mechanizations, 53-54
  miniature, 177
  vertical channel, 68-70
Information matrix, 93
Integrity. See GPS integrity
Integrity beacon, 129, 152, 159
  near-far problem, 129

K
Kalman, Rudolf E., 107
Kalman, filter
  intro, 93, 96-97
  computer needs, 102, 105, 142, 146
  divergence, 101, 195-199
  EKF, 102, 206
  erratic ramp/reset, 104
  loose coupling, 143
  measurement rejection, 104-105, 202
  initializing, 97, 104
  gain matrix, 99
  modeling, 104
  observability, 206
  ramp/reset discontinuity, 199
  simulations, 193
  tight coupling, 143
  tuning, 98, 143
Kinematic GPS, 151-152

L
Latitude
  astronomic, 22
  geodetic, 21-22
Least-squares problem, 91-93, 98, 121, 155
Longitude, 22-23
LORAN-C, 9

M
Matrix, meaning of, 16
Mean. See expected value; see also geometric mean
MIL-STD-1553B, 13, 142
MIT, xii, 93, 178-179
Monte Carlo simulation, 87, 106, 199
Most probable position, 73, 97
Multipath, 118, 135

N
Nl ambiguity, 152, 156
Navigation
  AHRS, 70
  definition, 1

230 INDEX

errors, 3
flight displays, 11-12
flight instruments, 42-45
flight testing, 106, 165-169
integrated, 3, 13, 132, 141, 168
outputs (PVAT), 25
slant range, 7
Navigation message. See GPS navigation message
NAVSTAR, See GPS

O
OMEGA, 9

P
P code, See GPS P-Code
Parkinson, Brad, 110, 138
Percentile error curves, 172-173
PDOP, 122
Pinson error model, 68, 77
Precession. See Gyro apparent precession
Precise Positioning Service, 112
Probabilistic event, 84
Probability Density Function, 85-86
Probability Distribution Function, 84-85
Pseudolite. See integrity beacon
 mobile pseudolite, 163
Pseudorandom noise, 110
Pseudorange. See GPS pseudorange measurements
Psi equations. See Inertial navigation error equations
PVAT, xi, 2

Q
Quaternions, 28, 66

R
RADAR, 10
RAIM, 158
Rate sensors. see gyro
Reference frames
 body, 2, 16
 computed, 63
 ECEF, 19, 115

Euler, 27
geocentric, 20, 23
inertial, 18
geodetic (tangent), 1-2
navigation, 20
platform, 57-58, 63
vehicle-carried vertical (geographic), 2
Regression analysis, 80
Relative bearing, 6
Residual, 116
Ring laser. See gyro ring laser
Root-mean-square (rms), 171
Root-sum-squared, 75, 158
Rotation matrix, 18. See also Vector cross-product

S
Sagnac effect, 51, 53
Satellites. See GPS satellites
Schmidt, Stanley, xii-xiii, 93, 101, 104, 207
Schuler frequency, 63, 168
Schuler tuning, 59-63
Selective availability, 132-134
Sensitivity analysis, 75
Sequential tracking receivers, 127
Sideslip, 44
Situational awareness, 2, 11-13, 101
Space integrator, see inertial navigation (geometric)
Spread spectrum. See GPS spread spectrum
Spectral density. See Stochastic process
Sperry, Lawrence, 14
Spilker, James, 125
Spoofing, 131
Sputnik, 137
Standard deviation, 85-86
Standard INS, 74
Standard rate turn, 44
Stanford University, 138, 159
State-space system, 78, 99, 104
State-vector, 79, 97-98
Statistics, 82
 circular error probable, 83-84, 173-174
 geometric mean data reduction, 171-173

Stochastic process, 87
    autocorrelation, 88
    bandwidth, statistical, 88
    covariance matrix, 89
    spectral density, 88
    stationary, 87
Strapdown INS. See inertial navigation, analytic strapdown
System 621B, 110

T

TACAN, 7
TDOP, 122
Time
    common view for GPS, 113
    MSL correction for GPS, 112
    GPS time, 112
    IRIG, 162-164
    Naval Observatory USNO, 19
    relativity correction for GPS, 112
    real time, 163-164
    TAI, UT0, UT1, UT2, 19
    ZULU, UTC, GMT 20
Time of arrival. See GPS pseudorange measurements
Trace of covariance matrix, 100
Transit program, 138

Translators, 162
Tropospheric delay, 118, 135, 145
Tunnel concept, 159
Turn-and-slip indicator, 44, 53

U

UERE. See GPS user equipment error

V

Variance, 86
VDOP, 111, 122, 143
Vector cross-product, 18
Vector differentiation. See Coriolis Theorem
Vertical Channel, 68-70
VOR, 6-7

W

Wander Angle, 63
Waypoints, 4
Wide-laning, 145, 156
Wiener, Norbert, 107

Y

Y code. See GPS Y-encrypted Code

TEXTS PUBLISHED IN THE AIAA EDUCATION SERIES

Integrated Navigation and Guidance Systems
Daniel J. Biezad 1999
ISBN 1-56347-291-0

Aircraft Handling Qualities
John Hodgkinson 1999
ISBN 1-56347-331-3

Performance, Stability, Dynamics, and Control of Airplanes
Bandu N. Pamadi 1998
ISBN 1-56347-222-8

Spacecraft Mission Design, Second Edition
Charles D. Brown 1998
ISBN 1-56347-262-7

Computational Flight Dynamics
Malcolm J. Abzug 1998
ISBN 1-56347-259-7

Space Vehicle Dynamics and Control
Bong Wie 1998
ISBN 1-56347-261-9

Introduction to Aircraft Flight Dynamics
Louis V. Schmidt 1998
ISBN 1-56347-226-0

Aerothermodynamics of Gas Turbine and Rocket Propulsion, Third Edition
Gordon C. Oates 1997
ISBN 1-56347-241-4

Advanced Dynamics
Shuh-Jing Ying 1997
ISBN 1-56347-224-4

Introduction to Aeronautics: A Design Perspective
Steven A. Brandt, Randall J. Stiles, 1997
John J. Bertin, and Ray Whitford
ISBN 1-56347-250-3

Introductory Aerodynamics and Hydrodynamics of Wings and Bodies: A Software-Based Approach
Frederick O. Smetana 1997
ISBN 1-56347-242-2

An Introduction to Aircraft Performance
Mario Asselin 1997
ISBN 1-56347-221-X

Orbital Mechanics, Second Edition
Vladimir A. Chobotov, Editor 1996
ISBN 1-56347-179-5

Thermal Structures for Aerospace Applications
Earl A. Thornton 1996
ISBN 1-56347-190-6

Structural Loads Analysis for Commercial Transport Aircraft: Theory and Practice
Ted L. Lomax 1996
ISBN 1-56347-114-0

Spacecraft Propulsion
Charles D. Brown 1996
ISBN 1-56347-128-0

Helicopter Flight Dynamics: The Theory and Application of Flying Qualities and Simulation Modeling
Gareth D. Padfield 1996
ISBN 1-56347-205-8

Flying Qualities and Flight Testing of the Airplane
Darrol Stinton 1996
ISBN 1-56347-117-5

Flight Performance of Aircraft
S. K. Ojha 1995
ISBN 1-56347-113-2

Operations Research Analysis in Test and Evaluation
Donald L. Giadrosich 1995
ISBN 1-56347-112-4

Radar and Laser Cross Section Engineering
David C. Jenn 1995
ISBN 1-56347-105-1

Introduction to the Control of Dynamic Systems
Frederick O. Smetana 1994
ISBN 1-56347-083-7

Tailless Aircraft in Theory and Practice
Karl Nickel and Michael Wohlfahrt 1994
ISBN 1-56347-094-2

Mathematical Methods in Defense Analyses, Second Edition
J. S. Przemieniecki 1994
ISBN 1-56347-092-6

Hypersonic Aerothermodynamics
John J. Bertin 1994
ISBN 1-56347-036-5

Hypersonic Airbreathing Propulsion
William H. Heiser and David T. Pratt 1994
ISBN 1-56347-035-7

Practical Intake Aerodynamic
Design
E. L. Goldsmith and J. Seddon,   1993
Editors
ISBN 1-56347-064-0

Acquisition of Defense Systems
J. S. Przemieniecki, Editor   1993
ISBN 1-56347-069-1

Dynamics of Atmospheric
Re-Entry
Frank J. Regan and Satya M.   1993
Anandakrishnan
ISBN 1-56347-048-9

Introduction to Dynamics and
Control of Flexible Structures
John L. Junkins and Youdan Kim   1993
ISBN 1-56347-054-3

Spacecraft Mission Design
Charles D. Brown   1992
ISBN 1-56347-041-1

Rotary Wing Structural Dynamics
and Aeroelasticity
Richard L. Bielawa   1992
ISBN 1-56347-031-4

Aircraft Design: A Conceptual Approach,
Second Edition
Daniel P. Raymer   1992
ISBN 0-930403-51-7

Optimization of Observation and
Control Processes
Veniamin V. Malyshev, Mihkail N.   1992
Krasilshikov, and Valeri I. Karlov
ISBN 1-56347-040-3

Nonlinear Analysis of Shell
Structures
Anthony N. Palazotto and   1992
Scott T. Dennis
ISBN 1-56347-033-0

Orbital Mechanics
Vladimir A. Chobotov, Editor   1991
ISBN 1-56347-007-1

Critical Technologies for
National Defense
Air Force Institute of Technology   1991
ISBN 1-56347-009-8

Space Vehicle Design
Michael D. Griffin and   1991
James R. French
ISBN 0-930403-90-8

Defense Analyses Software
J. S. Przemieniecki   1990
ISBN 0-930403-91-6

Inlets for Supersonic Missiles
John J. Mahoney   1990
ISBN 0-930403-79-7

Introduction to Mathematical
Methods in Defense Analyses
J. S. Przemieniecki   1990
ISBN 0-930403-71-1

Basic Helicopter Aerodynamics
J. Seddon   1990
ISBN 0-930403-67-3

Aircraft Propulsion Systems
Technology and Design
Gordon C. Oates, Editor   1989
ISBN 0-930403-24-X

Boundary Layers
A. D. Young   1989
ISBN 0-930403-57-6

Aircraft Design: A Conceptual
Approach
Daniel P. Raymer   1989
ISBN 0-930403-51-7

Gust Loads on Aircraft: Concepts and
Applications
Frederic M. Hoblit   1988
ISBN 0-930403-45-2

Aircraft Landing Gear Design:
Principles and Practices
Norman S. Currey   1988
ISBN 0-930403-41-X

Mechanical Reliability: Theory, Models
and Applications
B. S. Dhillon   1988
ISBN 0-930403-38-X

Re-Entry Aerodynamics
Wilbur L. Hankey   1988
ISBN 0-930403-33-9

Aerothermodynamics of Gas Turbine
and Rocket Propulsion,
Revised and Enlarged
Gordon C. Oates   1988
ISBN 0-930403-34-7

Advanced Classical Thermodynamics
George Emanuel   1987
ISBN 0-930403-28-2

Radar Electronic Warfare
August Golden Jr.   1987
ISBN 0-930403-22-3

An Introduction to the
Mathematics and Methods
of Astrodynamics
Richard H. Battin   1987
ISBN 0-930403-25-8

Aircraft Engine Design
*Jack D. Mattingly, William H. 1987
Heiser, and Daniel H. Daley*
ISBN 0-930403-23-1

Gasdynamics: Theory and
Applications
George Emanuel 1986
ISBN 0-930403-12-6

Composite Materials for Aircraft
Structures
Brian C. Hoskin and 1986
Alan A. Baker, Editors
ISBN 0-930403-11-8

Intake Aerodynamics
J. Seddon and E. L. Goldsmith 1985
ISBN 0-930403-03-7

The Fundamentals of Aircraft Combat
Survivability Analysis and Design
Robert E. Ball 1985
ISBN 0-930403-02-9

Aerothermodynamics of Aircraft
Engine Components
Gordon C. Oates, Editor 1985
ISBN 0-915928-97-3

Aerothermodynamics of
Gas Turbine and
Rocket Propulsion
Gordon C. Oates 1984
ISBN 0-915928-87-6

Re-Entry Vehicle Dynamics
Frank J. Regan 1984
ISBN 0-915928-78-7

Published by
American Institute of Aeronautics and Astronautics, Inc.
1801 Alexander Bell Drive, Reston, VA 20191